SHAMANS AND SHAMANISM

A COMPREHENSIVE BIBLIOGRAPHY

OF THE TERMS USE IN NORTH

AMERICA

BY

PETER N. JONES

Bäuu Press
PO Box 4445
Boulder, CO 80306
www.bauuinstitute.com

1. Shamanism. 2. North America. 3. Anthropology.

Shamans and Shamanism: A Comprehensive Bibliography
Of the Terms Use in North America
     / by   Peter N. Jones
          p. cm.

ISBN 13: 978-0-9820467-1-5

Bauu Press
Boulder, Colorado

Printed in the United States of America

INTRODUCTION

Shamanism... what is it? Is it a phenomenon with a clear definition or with a set of clearly definable attributes? Has the phenomenon changed over time, or are today's versions found in suburban basements the same as those that were practiced hundreds of years ago by various tribal peoples? What can we figure out about shamanism if we simply look at the term itself and how it has been employed over time? What if we restrict ourselves to one geographic location? These are some of the questions grappled with, and partially answered, in this book.

Ever since the phenomenon of shamanism was first reported in the late 19th century (Flaherty 1992; Hoffman 1888; Matthews 1888; Mikhailovskii 1895a; Mikhailovskii 1895b; Powell 1899) it has been of interest to scholars and laypeople alike. Though the phenomenon has existed for far longer, with some scholars arguing for its existence since the Paleolithic epoch (40,000-10,000 years before present) (Bednarik, et al. 1990; Greer and Greer 2004; Hedges 1983; Layton 2000; Lewis-Williams 2003a; Lewis-Williams 2003b; Wellmann 1981), few have conducted extensive qualitative research to determine whether the phenomenon existed historically, or exists today, within the specific geographic area of study. Likewise, rarely has it been asked whether one is simply appropriating a term (such as shamanism) from outside the specific area of inquiry and overlaying it onto observable phenomena within another area of inquiry.

Eliade (1951/1974) was the first to conduct a large cross-cultural comparative survey of the phenomenon, finding the phenomenon of shamanism to exist in six of the nine continents of the world. Over time,

1

this research has been criticized by many (see, for example, Kehoe 2000) because it took an etic stance to the phenomenon, defining shamanism on the reductionistic criteria of certain observable behavioral traits and characteristics. To surmount this limitation other criteria have been used to define shamanism over the years, ranging from the necessity of ecstatic experiences (Lewis 1989) to "journeys to nonordinary reality" (Harner 1990, 2002), to the designation of a complex set of attributes within specific hierarchically delineated sociocultural settings (Winkelman 1984a, 1989, 2000). None of these subsequent definitions, however, questioned whether the actual phenomenon of shamanism existed historically, or exists presently, within the area of study under discussion. Rather, these studies began like Eliade with the *a priori* assumption that the phenomenon of shamanism does (or did) exist, and that there was an agreed-upon definition of the phenomenon that was widely applicable and usable for comparative purposes. Thus, it was thought only necessary to simply look for particular behavioral traits—or whatever defining criteria were chose—within a specific geographic region in order to establish whether the phenomenon of shamanism existed or not.

This lack of a clear delineation of shamanism for scientific purposes has proven especially burdensome in light of the postmodern and poststructural critiques of cultural theory (Abu-Lughod 1991; Barnard 2000; Crapanzano 1992; Gebhardt 2002; Gellner 1992; Orta 2004; Ortner 1984; Wallerstein 2003). Likewise, with the explosion of the complimentary and alternative medicine (CAM) field within the last 20 years, and with the recent attempt to incorporate so-called shamanic techniques or modalities into contemporary health care practices (see, for example Cardena 1999; Erdmann 2003; Ingerman 2003; Money 2000; Porterfield 1984; Winkelman 2003; Zatzick and Johnson 1997), the need for a clear understanding of the phenomenon is even more essential today. Finally, the study of shamanism has been of critical interest not only for anthropologists, but also for sociologists, historians, psychologists, and religious studies scholars since it was first recognized in the 19th century (Flaherty 1992; Narby and Huxley 2001), and if any interdisciplinary dialogue and development in our understanding of the term and its associated phenomena is to occur then a clear understanding of the phenomenon is essential.

As a step towards remedying this increasingly confusing situation, this book examines the phenomenon and term of shamanism from what I call a concept-based geographical approach. That is, in this book I look at the phenomenon as a concept and how that concept has been employed within a specific geographical area—namely North America. This book, therefore, is not concerned with attempts at discovering the ontogenetic history of the term *shaman* itself, which would require a much larger research question, one that would focus more on northern Asia and eastern Europe. Some preliminary work on this question has already been done on this question by Hoppal (1989) and Langdon (1989). The two primary purposes of this book are, rather: (1) to discuss the historical use of the terms *shamanism* and *shaman* in North America; and (2) to compare current understandings and descriptions of the phenomenon with those of the historical and archival record.

In the process of researching this book, a comprehensive bibliography was created, which has been added at the back. This bibliography is quite informative in its own right, as one is able to see the changes in use of the terms over the course of the last hundred years. Furthermore, over 780 references are included in the bibliography, making it an important resource for other investigations of the phenomenon and terms *shamanism* and *shaman*.

Peter N. Jones

DEFINITIONS, TERMS, AND THEMES USED

Prior to discussing the term and phenomenon of shamanism as it has, and continues, to occur in North America, it is important to clearly define several key concepts. First, this book and its discussion is limited to the northern portion of the North American continent, which is defined for the purposes of this book by current sociopolitical criteria. Therefore, "North America" covers the area within the current nation-states of Canada and the United States but does not include Mexico, which some scholars include within North America (as opposed to delineating a separate Central or Middle America). This defining characteristic is in line with the original conceptual delineation of "culture areas" by Kroeber (1939), Wissler (1914), and other anthropologists, although it does allow for the possibility of excluding examples and data from some groups that may have prehistorically and historically inhabited both sides of the U.S.-Mexico border (for example, various Apache Native American groups).

Along with the geographic definition that helped guide the research and preparation of this book, several corresponding themes were also addressed: occurrence, prevalence, and form.

5

<u>Themes Addressed</u>

Occurrence

Prevalence

Form

Occurrence was the first theme addressed in the preparation of this book. This is important, for example, because according to some researchers (Axtell and Aragon 1997; Downton 1989; Grossinger 1990; Hultkrantz 1997; Lewis-Williams 2003b; McClenon 1997; Riches 1994; Samuel 1996; von Keitz 1999; Winkelman 1984a, 2000, 2002), the practice of shamanism was historically quite prevalent in North America, but it is currently found only among traditional American Indian and First Nation peoples, though some do allow for the occurrence of various shamanic *techniques* to exist and be in use in contemporary society by non-American Indian peoples. As such, the occurrence of the phenomenon, and the particular groups believed to be associated with the phenomenon, was an important guiding theme. As will be discussed in the conclusion, the phenomenon and use of the term are no longer able to be claimed by any one group or culture, as its use is currently found among disparate cultural, social, economic, ethnic, and ideological groups.

Along with the theme of occurrence, and as a partial result of it, the theme of prevalence was also noted. If shamanism does (or did) occur in particular areas of North America, it was important for the purposes of this book to ask how much, and to what level. The reason is that many researchers have implied that shamanism was quite common in indigenous societies throughout the Americas, Europe, Africa, and Asia, and that it was in fact a phenomenon practiced by many individuals (Eliade 1951/1974; Grossinger 1990; Harner 2002; Krippner 2002; McClenon 1997; Money 2001; Noll 1985; Ripinsky-Naxon 1993; Winkelman 1995, 2000, 2002, 2004). Furthermore, this theme not only addressed the issue of how much, but also the important question of who, for many scholars have also argued that shamanism consists of a set of techniques that are biologically wired into all humans, and that only some have "unlocked" the secrets of these techniques (d'Aquili and Newberg 1998, 1999, 2000; Krippner 2000; Krippner and Combs 2002a; Krippner and Combs 2002b; Peters 2001; Winkelman 2000, 2002, 2004). If shown to be true, then this would mean that so-called shamanic techniques are available to everyone and that the knowledge associated with the phenomenon is available to everyone. As discussed at the end, because the use of the terms shamanism and shaman have changed over time, the occurrence of the phenomenon and the individuals who practice and participate in the phenomenon have also changed.

Finally, the theme of form was addressed. To balance the line of inquiry across the themes of occurrence and prevalence, it was necessary to ask how the phenomenon of shamanism manifests itself (or was manifested) in North America, and whether it was different in the past then now and whether it manifests differently in different areas. Winkelman (1984a, 1991, 1995, 2000, 2002, 2004), for example, has argued that shamans only occur in societies that are either hunter-gatherer based in their subsistence regime, or who perhaps practice (or practiced) pastoral subsistence regimes. If this is correct, then one would not expect to find shamanism in scholarly literature dealing with contemporary North America, but instead, to be a phenomenon of the purely historical circumstances of pre-contact American Indian and First Nation peoples. Again, as I will discuss at the end, the form of the phenomenon has greatly changed over the last hundred years, and is particularly evident in the bibliography at the back of this book.

Beyond these few guiding definitions and themes, the data and methods utilized in the research and preparation of this book were left open. As noted, one of the purposes of researching this book was to investigate the history and uses of the terms shaman and shamanism. If any further constraints were placed on the research agenda, such as limiting the scope of the project to a single academic field, then the possibility of a bias in the data and conclusions presented would emerge. As can be inferred from an examination of the comprehensive bibliography presented in this book, the terms shaman and shamanism have been used in an array of fields to discuss numerous, and potentially mutually exclusive, phenomena.

HISTORY AND
USE OF THE
TERMS

Based on the data gathered for this book, the first uses of the terms in North America occurred in the middle to late 1800s. the first recorded example is found in the Annual Report of E.M. Gibson from Washington Territory to the Commissioner of Indin Affairs in 1872. Evidence indicates that this and other 19th century records that used the terms *shamanism* and *shaman* in North America were written from the point of view of the interested, yet disbelieving, Western European or Euroamerican. Whether discussing shamans in the Plateau, Great Basin, Eastern Woodlands, or the Arctic culture regions, the early reports attempted to come to terms with the exotic "otherness" of the phenomenon being reported.

For example, most of the reports discussed both the phenomenon's acoustic and visual aspects, treating it as a gigantic theatrical performance. Likewise, the shaman's "costume" was noted in detail, as well as any other objects or paraphernalia used within this seemingly showmanship-based event. As Flaherty has discussed, however,

Greatest attention was usually given to the trance state: not only to attaining it and recovering from it, but especially to its genuineness. Again and again, the reports contained phrases like "dead to the world," "as if dead," and "paroxysm of unconsciousness and powerlessness." Those who lived in an age that feared apparent death—*Scheintod*, that condition in which the body's vital signs are negligible or seem nonexistent—could hardly begin to conceive of such an imbalanced mind-body relationship. (1992:10-11)

In this way, early understandings of the phenomenon in North America conceived of it as a large theatrical performance that involved a great amount of audio and visual manipulation. This theatrical performance, often observed or accompanied by several community members, was mainly centered on the individual who could achieve the miraculous, namely becoming "dead" (for example, see Hoffman 1888; Matthews 1888; Smith 1896). During this time, the term *shaman* was also used interchangeably with such terms as *pretres, pretresses, mages, devins, hierophants, druides, medicins, saiotkatta, agotsinnachen, arendiouannens, agotkon, piayes, boyes, pages, charlatans*, and, most often, *jongleurs* (Flaherty 1992; Narby and Huxley 2001).

Because multiple terms were used in early reports to refer to the phenomenon of "shamanism," when anthropologists began to explicitly use the term *shaman* within the academic literature, a similarly wide concept of the phenomenon was employed. At this time, anthropology was dominated by the theories of cultural evolution developed by Frazer (1890/1993), Morgan (1877/1985), Tylor (1924/1871, 1924/1881), and other so-called armchair anthropologists. The theories they developed were based on the early reports of missionaries, fur trappers, ethnologists, and others who had reported on the exotic "otherness" of the people they encountered. For example, so-called "primitive" traits, which were discussed in these early reports, were taken to represent survival mechanisms from archaic times, while the question of the origins and evolution of human culture dominated much of the discussions.

Starting from the mid-nineteenth century, observers began to describe these individuals by the Siberian term "shaman." This identification was misleading, in that the common term ignored regional and tribal differences between religious leaders across North America, while North

American "shamans" differed from the precise phenomenal pattern well known from Siberia. Furthermore, modern-day Native activists object to using the all-purpose term "shamanism" in the North American context. But in the nineteenth century, speaking of shamans marked real progress from the older analogy of the medicine man as deceptive Catholic priest. It established Native religion in a global context, appropriate to a particular stage of human development.

As a result of the development of this concept, ethnographers were forced to explore more accurately what these individuals actually did, and in the process they found that these practitioners commanded impressive skills. For example, in 1894, Washington Matthews wrote that "the accomplished Navajo shaman must be a man of superior memory and of great intellectual industry." However dubiously these individuals were regarded, they were now more closely studied and their religious system much better understood (and respected). For example, a 1904 exhibition at the American Museum of Natural History collected shamanistic materials from Asia, as well as North and South America, in order to portray Shamanism (capitalized) as an archaic global faith. Reporting the exhibition, the *New York Times*' headline read "Ancient Religion of Shamanism Flourishing Today;" the article remarking that the tradition was much older than Judaism. "It is perhaps the only remaining and enduring type of culture extending from the remotest, probably from Paleolithic, times, to date" (Maddox 1904).

University-trained anthropologists first reporting in the scholarly literature on "shamanism" likewise considered it an example of the earliest form of religion, that designated by Tylor (1871/1924) as animism, primarily because so-called shamanic ideology included the belief in souls of inanimate objects, animate "beasts," and the ability to become "dead." While shamanism was treated as an example of an animistic religion, shamanic techniques and practices were separately analyzed as acts of magic. As is well known, the first to deal with magic systematically within the anthropological literature was James Frazer (1890/1993). According to Frazer, magic was separate from religion in that its purpose was not directed at worship but rather at altering events. Frazer called magic a pseudo-science in which the "primitive" (i.e., as opposed to the "civilized") had a logical, but also faulty, perception of reality. Furthermore, because Frazer claimed that magic was based upon the "laws of sympathy," magic and the magical techniques employed by "shamans" were really just technical acts that had the goal of altering the world. In reality it was argued that these techniques were inefficacious, since acts based on the law of similarity were thought to have no causal power.[3] The terms *shamanism* and *shaman*, as the comprehensive bibliography corroborates, were employed within this discourse to describe theatrical events that included magical acts and techniques grounded in an animistic ideology (see, for example, Powell 1899; Smith 1896). Furthermore, the terms were considered to define a phenomenon found only in so-called "primitive" cultures, such as among Native American and First Nation peoples (as they were culturally framed at this time), because Frazer's evolutionary theory argued that true science would replace magic as the "primitive" moved up the developmental scale to becoming "civilized."

With the beginning of the 20th century, this unilineal theory of mental and cultural evolution was appropriately rejected for its ethnocentrism, along with its inability to be replicated by empirical evidence. In North America, anthropologists subsequently adopted what became known as a historically particular paradigm, applying this new theoretical orientation to the study of the phenomenon described by the terms *shamanism* and *shaman*. Historical particularism was a theoretical orientation developed by Franz Boas (1896/1948, 1911, 1930, 1933), who took Tylor's redefinition of the German word *Kulture*,[4] and used it in search of natural laws that governed the *state* or *condition* of cultural phenomena (rather than the phenomena themselves). Although Boas originally defended a strictly geographical particularist viewpoint in this research agenda, he would later broaden the particularist viewpoint in his search for nomothetical laws. Thus, the term *culture* within the American school of anthropology came to mean a state or condition that could be delineated and investigated through the four-field approach of ethnology (which would later become ethnography and cultural anthropology), archaeology, human remains (later to be called physical anthropology), and language (later to be conjoined with linguistics). Boas argued that this four-field approach was the proper approach for arriving at the state or condition of any cultural phenomenon (e.g., shamanism).

> When we have cleared up the history of a single culture and understand the effects of environment and the psychological conditions that are reflected in it we have made a step forward, as we can then investigate in how far the same causes or other causes were at work in the development of other cultures. (Boas 1948: 279)

12

Boas would later revise this version of historical particularism, arguing that there may not necessarily be nomothetical laws governing cultures that can be comparatively analyzed, but instead that the occurrence of similar *institutions* throughout the world reflected something inherent in the human mind (Boas 1911). It is this later reformulation of historical particularism that allowed the terms *shamanism* and *shaman* to be used cross-culturally and cross-geographically, as they described an institutional phenomenon removed from its specific sociocultural ontogeny. At the same time, Tylor's and Frazer's theories about the early formulations of religion and magic were also being incorporated into developing anthropological theories. This further strengthened the view in North America that the terms *shamanism* and *shaman* did not just define a theatrical performance but that they also defined a *function* and *structure* of a cultural institution under the historical particularist viewpoint (see, for example, Radin 1911; Speck 1919).

As is well known, the early 20th century was also the period when behaviorism arose within psychology, and functionalism arose within philosophy, anthropology, and sociology. The latter of the two, functionalism, appears to have had a profound effect on the definitions of these terms at this time in North America. For example, Durkheim's functionalism succeeded evolutionism, and this theoretical orientation had a pervasive effect on North American scholars using the terms. Though not an evolutionist, Durkheim continued the inquiry into the origins of culture and religion in *The Elementary Forms of Religious Life* (1915). In this work, he concluded that religious thought emerged from the social structure and that religious rites maintained the social structure's order. Further, Durkheim placed magic within the private realm, effectively minimizing his discussion of it in his theory of religion, although this further theoretical refinement solidified the functional component recently added to the understanding of the terms *shamanism* and *shaman*. Shortly thereafter Mauss (1902/1999) drew from Durkheim's thesis of religion in order to create a theory of magic, which also had a simultaneous effect on the definitions and uses of the terms *shamanism* and *shaman*. Concerned, as were his predecessors, with constructing universal and objective (that is, operational) definitions, Mauss accepted Durkheim's idea that religion was a public act, separate from magic, which he argued was a private act. Mauss further argued that while religion created an ideal or moral order for society, magic tended towards "evil" and to individual goals. Influenced by Mauss' theories on magic, *shamanism* and *shaman* came to be used to refer to secret

magical acts performed by individuals that were based in a tradition trans-
mitted and believed in by a group. That is, Mauss' theories led to the argu-
ment that the origin of the phenomenon described by the term *shamanism*
was social, as was religion's, but that it represented the individual's
(shaman's) response to the collective.

Mauss' theories, along with the general trend for evolutionary
approaches, further altered the definitions of *shamanism* and *shaman* by
relying on functional characteristics to define the phenomenon. For example, Mauss contended that magic became less important the more civilized
a culture became because its function lessened, a theory that allowed scholars to limit the phenomenon and the use of the terms to people thought to
be "primitive" in terms of cultural or psychological development. Because
shamanism and *shaman* were already associated with the so-called "primitive," Mauss' theoretical arguments further limited the applicability of the
terms to descriptions of a phenomenon only taking place in so-called
"primitive" groups (i.e., American Indian and First Nation peoples as they
were framed within the social sciences at this time).

Thus, by the early 20th century, the terms *shamanism* and *shaman*
had shifted from defining a type of theatrical performance that involved
large amounts of audio and visual sensations centered on an individual who
could achieve the miraculous (Flaherty 1992; Flaherty and Fink 1995;
Hoffman 1888; Matthews 1888; Smith 1896), to an individual who practiced
a form of magic for personal gain, but who participated within, and upheld,
the social order (*sensu* Durkheim 1915; Mauss 1902/1999). This new definition and application of the terms are represented by such works as that of
Dixon (1904, 1908) and Radin (1911, 1914), the latter of whom discussed
the *function* of shamans in maintaining the social order of the Winnebago
American Indian people. In conjunction with this redefinition process, and
as a result of Boas's historical particularism, the phenomenon defined by
these terms further came to be considered a cultural institution among so-
called "primitive" peoples; one that could be compared with similar cultural institutions around the world, as exemplified in the work of Wissler
(1916), Benedict (1922), and Park (1934). Similarly, archaeologists at this
time began to use this understanding of the term shamanism. "When Lowry
Ruin was excavated, archaeologists found secret passages in kivas, which
allowed figures to make seemingly supernatural entrances during rituals.
The archaeologist concerned was quoted as saying that 'the shamans had to
make a living and to do that they had to fool the people'" (Jenkins 2004: 32).

As can be seen by examining the comprehensive bibliography included at the back of this book, the use of *shamanism* and *shaman* by the early 20th century had already come under reductionistic and etically bound definitional criteria, radically altering the definition of the terms from the previous generations, but also from the actual phenomenon as practiced and understood within each particular culture. In the attempt to construct mutually exclusive categories of religion and magic, the defining criteria for the terms *shamanism* and *shaman* were reduced to observable characteristics thought to reflect private magical acts that maintained the social order. As the comprehensive bibliography indicates, the next theory to have an effect on the definition of *shamanism* and *shaman* in North America and continuing the previous generation's reductionistic tendencies was functionalism, as largely developed by Bronislaw Malinowski.

Malinowski's functionalism was based on ideas of so-called human psychological needs (i.e., functions), which were analyzed in terms of the questions raised by Tylor, Frazer, and others regarding magic and religion. Unlike earlier theoretical attempts, he negated the spurious conclusions that had arisen due to the idea of a "primitive" mentality. Instead, Malinowski created the Eurocentrically oriented category of a "savage," who was thought to have the same psychological capacities and cultural characteristics as found in so-called "civilized" individuals, but who believed in and practiced magic. Under this theoretical framework, the reason for magic's existence lay rooted in areas of uncertainty in the human experience and not in fallacious perceptions of the world and the lack of a rational, empirical science as previously thought (Malinowski 1927, 1944, 1948).

Malinowski theorized that magic was a function that did not work in altering events but instead existed because it performed a very important *function* for the human psyche, namely, relieving anxiety. As an individual's or culture's cognitive abilities evolved (towards a Eurocentric ideal)—and improved the individual's or culture's control over nature—Malinowski contended that magic would die out or become less important, only existing in the gray areas of experience that escaped control. Similarly, *shamanism* and *shaman* could only be used when discussing phenomena associated with the so-called "savages," because at this time they referred to a phenomenon based in magic. In North America, this meant that the term was only used in reference to discussions concerning Native American and First Nation peoples. As Shamdasani (2003) has strongly argued,

15

Malinowski's theories had a profound impact on the thought of Carl Jung, especially the function magic was thought to have in alleviating human anxiety. The work of Jung, along with the functionalism of Malinowski, in turn, had a large impact on anthropologists and psychologists in North America who began collaborating at this time. For example, it was Jung's theories that many humanistic and transpersonal psychologists used when they began to discuss *shamanism* and *shaman* in terms of psychological characteristics (see, for example, Boyer 1964b; Gunn 1966; King 1960; Murphy 1964; Olson 1961; Posinsky 1965).

Prior to humanistic and transpersonal psychology's use of the terms, however, psychologists used the terms mostly in the pop literature, primarily because at this time academic psychology was centered in the laboratory and in generating experimental data (Leahey 1987). The theories of Durkheim, Boas, and Malinowski all contributed to the early discourse and understanding of the terms *shamanism* and *shaman* because these scholars were all students of the German psychologist Wilhelm Wundt, a pioneer in the field of folk psychology and mental capacities. Wundt attempted to find psychological explanations to beliefs and actions through the use of facts supplied by ethnology. He contrasted psychological stages, such as the "totemic stage," to the "age of heroes and gods," to the "enlightened age of humanity," and associated each with a distinctive type of cognitive thinking (Wundt 1896). Unlike other theorists of the time, he believed that "primitive" and "civilized" humans had the same intellectual capabilities, they just exercised them differently.

The supposed cognitive differences of individuals in contrasting cultures were not, however, part of the goals of the emerging discipline of anthropology at the beginning of the 20th century. As was noted above, for example, Franz Boas was opposed to the seeming reductionism of psychology, which he argued minimized complex historical phenomena into a few basic ideas. He was also strongly opposed to the racism that went along with much of the psychologically influenced thought of the time. In *The Mind of Primitive Man* (1911), Boas noted that if anthropologists could show that mental processes among so-called "primitive" and "civilized" people were essentially the same, the view that the different cultures existed in different stages of an evolutionary series could not be maintained. For Boas *shamanism* and *shaman* defined historically particular social institutions that could be compared to other institutions cross-culturally, and that the mentality or cognitive abilities of the "shamans" never came under direct investigation by anthropologists at the time. It was not until the 1920s, in fact, that a more psychological approach to studying culture and its phenomena, including the phenomenon defined by the terms *shamanism* and *shaman*, moved into the forefront in North America.

This movement was largely the result of the formation of the Culture and Personality approach, founded primarily by Edward Sapir, Ruth Benedict, and Margaret Mead. In 1934, Sapir posited that "the more one tries to understand a culture, the more it seems to take on the characteristics of a personality organization" (Sapir 1949: 201). He went on to assert that patterns of culture are connected by "symbolism or implication," and that an ethnographer must get beyond the superficial categories usually explored, such as kinship, ritual, or subsistence, to fully understand the connections that make up these patterns. Thus, he encouraged anthropologists to focus their studies on the individuals within specific cultures, because he argued that individuals look for and create meaning in their world, acting as a microcosm of the culture in which they live. Sapir's ideas became the foundation on which Culture and Personality theory was built (Bock 1994; Hsu 1972; Mandelbaum 1949; Sapir 1949).

This reformulation had a profound effect on the study of the phenomena defined by the terms *shamanism* and *shaman* within North America, for instead of defining the terms as cultural institutions that could be studied along side other cultural institutions, Sapir's formulation of the Culture and Personality theory allowed for scholars to define them based on single individuals. For example, at this time, the literature becomes replete with single case studies of Native American individuals who are termed *shamans* (for example, Johnson 1943; Kelly 1936, 1939; Park 1934, 1938; Ray 1936). These case studies, as was the goal of the Culture and Personality theory, characterized these individuals as reflecting particular aspects of their culture, focusing on these individuals, their behavior, and what that reflected of the culture. These characteristics are corroborated by the fact that psychological concepts of basic personality structure and modal personality were relied upon by the Culture and Personality theorists. For example, psychoanalyst Abram Kardiner and anthropologist Ralph Linton divided culture into what they called *primary institutions* (such as the institutions of subsistence or child training), which they thought produced a common denominator of basic personality. This basic personality could then theoretically be translated into secondary institutions such as religion, ritual, and "shamanism."

The notion of basic personality structures placed cultural integration as the focal point, acting as a common denominator of the personalities of the individuals who participated in the culture. Furthermore, it attempted to comprehend the causal relationship between culture and personality, rather than just taking it for granted, as cultural patterning had done. As such, individual cultures were thought to be integrated because all the members of a society were thought to share experiences that produced a basic personality structure, which in turn created and sustained other aspects of the culture. Under this theoretical orientation, the terms *shamanism* and *shaman* came to define what was thought to be an essential component of the culture under study because this phenomenon was thought to help integrate disperse aspects of that culture. As a result, however, scholars continued to focus more on the behavioral characteristics and etically observable aspects of "shamanism," which could be discussed in terms of functional or structural components of integration. The observable characteristics used to define these terms in North America, therefore, over time came to be classified as *techniques* that functioned to produce an integrated personality structure of the culture. This approach, however, shifted the defining characteristics of the phenomenon away from its theatric and historically particular context, to characteristics defined more on functions, behaviors, and individual techniques.

As various empirical incongruities were encountered in basic Culture and Personality orientations, the concept of modal personality was conceived of by Cora Du Bois (1944). Psychological anthropologists and social psychologists found this approach more acceptable than basic personality structure because it expressed the *character* (i.e., stereotypes) of a group as being the most frequent type encountered, rather than generalizing that traits were absolute or that all of the members of a culture had the same personality structure. As can be seen by examining the comprehensive bibliography, these Culture and Personality orientations had profound effects on the uses and definitions of the terms "shamanism" and "shaman" within North America. For example, in Honigmann's (1949) article, he looks for parallels in the development of "shamanism" between northern and southern Athapaskan American Indians, trying to find universal characteristics between the modal personalities of the "shamans." Furthermore, the comprehensive bibliography indicates that the Culture and Personality school also largely reduced the defining criteria of the phenomenon to a universal modal personality *trait*, one that could be compared to other cultural traits.

At the same time, other theoretical orientations stemming from the field of psychology also appear to have contributed to the reduction in characteristics used to define "shamanism" and "shaman" in North America. For example, projective tests, taken from experimental psychology, were being used to study personality in non-Western cultures and included the Rorschach Test and the Thematic Apperception Test. These tests were intended to let the subjects project themselves into a situation that was thought to give the ethnographer a window into the person's psyche. Thus, Boyer (1964a) used a Rorschach Test to study Apache "shaman's" personality as compared to non-shamans. The use of these methods, however, came under fire because of the subjectivity of their scoring and the emphasis on Occidental (i.e., Western) interpretations. The Rorschach Test, therefore, was used not to distinguish "shamans" from non-shamans, but in fact to compare personalities between two etically designated cultural institutions superimposed onto emic distinctions between individuals within a culture.

By the 1950s the Culture and Personality approach began to decline in popularity. Critics attacked the Culture and Personality approach for being untestable and unverifiable. Furthermore, the assumption that each culture could be characterized in terms of a single personality type allowed cultures to be oversimplified, leading to the spurious deduction that individuals within the culture were uniform and integrated (Bock 1988; Bock 1994). Likewise, it was impossible for scholars to describe the psychological characteristics of a culture without expressing their Western biases, a point supported by the comprehensive bibliography (see, for example, Honigmann 1949; Opler 1946; Smith 1954; Titiev 1956). However, this did not apparently concern Eliade (1951/1974), who at this time was writing his seminal book on the phenomenon, a book largely grounded in the comparative method as formulated in the ideas of the Culture and Personality theorists (Eliade 1953/1968). As Eliade noted in his foreword to *Shamanism: Archaic Techniques of Ecstasy*:

> As we have said more than once elsewhere, and as we shall have occasion to show more fully in the complementary volume (in preparation) to *Patterns in Comparative Religion*, although the historical conditions are extremely important in a religious phenomenon...they do not wholly exhaust it. (1951/1974: xiv)

He goes on to argue that an etic analysis using the comparative method is essential in coming to understand a phenomenon, a theoretical idea grounded in Culture and Personality theory.

> For, as we have attempted to show in *Patterns in Comparative Religion*, the very dialectic of the sacred tends indefinitely to repeat a series of archetypes, so that a hierophany realized at a certain "historical moment" is structurally equivalent to a hierophany a thousand years earlier or later. (Eliade 1951/1974: xvii)

As Kehoe (2000) has argued, one of the principle limitations in Eliade's book, and of this cross-cultural comparative approach, is the process of reducing and etically defining the phenomenon encompased by the terms *shamanism* and *shaman* into universal "personality" traits. This process erroniously circumscribes the very phenomenon when examined emically within each specific culture.

At the same time Frank Waters published his now famous book *Masked Gods* (1950). The title Masked Gods was meant to suggest the actual masks of figures like the kachinas, but also the concealed personality traits that emerged during psychoanalysis. Waters was fascinated by developments in contemporary psychiatry, psychology, and anthropology and found linkages between the buried forces in the unconscious and the mystic powers claimed by occult writers. His speculations gained support from the academic study of American Indian mysticism by reputable theorists who were influenced by psychoanalysis, as well as by psychosomatic interpretations of disease. These accounts praised exactly those features of indigenous religion and spirituality that nineteenth-century observers had condemned as irredeemably primitive: the works of "shamans." Some suggested that the music and chanting that characterized American Indian rituals might actually contribute to curing psychosomatic illnesses, while perhaps American Indians had pioneered modern discoveries in the medical uses of hypnosis. In 1936, an article in the *American Journal of Psychiatry* examined "Some points of comparison and contrast between the treatment of functional disorders by Apache shamans and modern psychiatric practice" (Opler 1936: 1371). In 1941, *Psychiatry* reported a study of Navajo religion by Alexander and Dorothea Leighton, who praised the effectiveness of traditional healing rituals, with their emphasis on group and community support, and on shared emotional experience. When a "shaman" sucked at a body and drew forth a worm or a thorn, this was considered a positive psychosomatic effect on the patient, who would see the cause of the illness visibly removed from their body (Leighton and Leighton 1941).

After the criticism of the 1950s, psychological anthropologists and social psychologists turned towards what came to be called the cross-cultural correlational approach, largely the product of psychologists and anthropologists at Yale University, including Clark Hull, John Dollard, G. P. Murdock, and John Whiting. These scholars believed that previous Culture and Personality theorists used their hypotheses as proven theories, which they then subsequently implemented to interpret specific case materials rather than test the hypotheses themselves. In cross-cultural correlational testing, on the other hand, a condition was looked for as it occurred or failed to occur in a culture. Then a condition considered related to the first was documented as present or absent in the other culture. The practitioners of this method believed that it was possible to determine whether there was a consistent correlation between the two conditions, and that this would confirm or negate the hypothesis. However, this method suffered from the same problem as the earlier Culture and Personality approaches, in that it was impossible to demonstrate that objective individuals could be found to judge the degree to which a culture had a certain custom or trait. It also necessitated drawing strict, and somewhat arbitrary, boundaries for cultures and the phenomena encountered in each culture.

This correlational testing method, however, remained somewhat popular throughout the 60s, 70s, and 80s, until it was largely abandoned by anthropologists and social psychologists as a result of poststructural and postmodern critiques. However, this method has had a large impact on the uses and definitions of the terms *shamanism* and *shaman* in North America. For example, Winkelman (1982, 1984b, 1989, 1991, 1996, 2000, 2002, 2004) has based his work on correlational testing methods, largely relying on data extracted through the Human Relations Area Files in the 1980s. For Winkelman and others using the correlational method, the phenomenon defined by these terms is reduced to a series of behavioral traits that are thought to hold empirically valid categorical positions, allowing them to be cross-culturally compared. Through this method of testing and definition making, phenomena are reduced to characteristics or techniques that can then be examined across different groups or datasets, taking out the cultural component of the phenomenon under study.

Furthermore, as a result of reducing the traits used to define these terms for correlational testing purposes to techniques or characteristics, the phenomena associated with these terms also became divorced from the larger emic phenomenon as a whole. Thus, in recent years, the use of the terms in North America have become replete with such studies as the psychophysiological effects of so-called "shamanic" drumming (Lane, et al. 1998; Maurer, et al. 1997; Neher 1962; Winkelman 2003), so-called "shamanic" mental imagery cultivation (Money 2000; Money 2001; Noll 1985), and the benefits of psychotropic plants that "shamans" are supposed to ingest (Bravo and Grob 1989; Dobkin de Rios and Winkelman 1989; Riba, et al. 2001; Schultes 1998; Wellmann 1981; Wiercinski 1989; Winkelman 1995). However, all of these studies fail to consider the history of the terms *shamanism* and *shaman* as discussed herein, and instead rely on the definitions put forth by the Culture and Personality theorists and the cross-cultural correlational theorists.

Defining *shamanism* and *shaman* based solely on characteristics or techniques that have been removed from any larger cultural context culminated in the 1980s with the publication of Michael Harner's book *The Way of the Shaman*. This book inspired a new interest in shamanism not just as something interesting that "they" (Native peoples) did, but as something common, everyday people could also explore. Harner himself wrote not just as an academic anthropologist, but as "an authentic white shaman."

The current study of the phenomenon defined by these terms consequently has become extremely loose and nonsensical. As this book argues, the uses of the terms *shamanism* and *shaman* have radically changed over the last 150 years within North America. This radical change in the definition of the phenomenon began with the understanding that "shamanism" described a theatrical performance and that the "shaman" was the performer, both of which were part of a specific cultural context. Subsequently, the phenomenon came to describe a social institution thought to functionally uphold the sociocultural order that it was part of, to the present understanding based on a complex set of specific techniques and characteristics that are thought to be free of any larger social or cultural context.

As a result, contemporary uses of the terms cover the entire spectrum, often referring to particular techniques or methods of behavior that are associated with the phenomenon. As discussed above, this is a logical consequence of the terms' historical uses. Common examples of these techniques and methods include such things as "shamanic soul retrieval," "heart vision shamanic journeys," and "animal spirit connections." For example, during a so-called "heart vision shamanic journey" one hypothetically contacts their animal helpers, ancestors, guardian angels, or spirit guides. Likewise, the individual supposedly reclaims their power, resolves trauma or past-life issues, or retrieves lost soul qualities. These techniques and characteristics are free of any direct social or cultural context, and rather rely on what is referred to as the participants own "personal mythology."

Furthermore, there are now numerous schools and institutions that claim to offer certification in one or more of these so-called "shamanic" techniques or modalities. In 2003, for example, Bill Brunton held a workshop in Minneapolis-St. Paul on "The Way of the Shaman." Brunton is "an anthropologist, shamanic practitioner and a member of the faculty of the Foundation for Shamanic Studies." Participants would "be initiated into shamanic journey, aided by drumming and movement techniques that induce a shamanic state of consciousness and can awaken dormant spiritual abilities and connections with Nature." The foundation offers its "rigorous training" to five thousand clients each year. Other courses by the foundation in 2003 included "Five-Day Soul Retrieval Training with Sandra Ingerman" in Santa Fe, while San Francisco was host to "Five-Day Harner Method Shamanic Counseling Training." Those registering for soul retrieval were warned that before registering, "please make sure that you have been having success in contacting your power animals and/or teachers on your own and that you feel confident about your journey skills." Another workshop "will introduce you to common shamanic practices; we will discuss and create sacred space, discuss shamanic practices, and journey to meet our power animals" (Jenkins 2004).

One of the most ambitious of these institutions offering training in shamanism is the Arizona-based Deer Tribe Metis Medicine Society, founded in 1986 by Harley Swift Deer Reagan. According to its website, tribal leaders "are modern day representatives of an ancient lineage of sacred knowledge of universal laws, ceremonial alchemy, healing techniques, alignment and communication with the elements of nature, magick, controlled dreaming, spiritual awakening and determination. Like the arcane mystery schools of the Sufi, the Druidic and Celtic, the Tibetan, and other Great Power traditions, the Sweet Medicine Sun Dance Path has evolved over thousands of years." The Deer Tribe traces its origins to the

Twisted Hairs Medicine Society, a "magickal mystery school," dating back in the Americas to 1250 BC. In the 1970s the group reportedly designated Swift Deer to bear its message to the modern world. Deer Tribe members participate in a wide range of activities, including rite of passage ceremonies, longhouse programs, seasonal and pipe ceremonies, healing rituals, and martial arts. Controversially, the group also advertises its "*Chuluaqui Quodoushka* spiritual sexuality teachings ... based on sacred shamanic traditions that integrate spirituality and sexuality into your life." Allegedly, these Q rituals represent an amalgam of Cherokee and Mayan traditions. Dozens of local Deer Tribe lodges and study groups are scattered across the United States and Canada.

Maine's Standing Bear Center for Shamanic Studies offers similarly varied activities, including shamanism, drumming, and smudging. A Florida program included an "Experiential Workshop of Smudging, Drumming, Shamanic Journeying, Medicine Wheel Building, and Crystal Grid Meditation and Fire Ceremonies." Along with these schools and institutions, numerous others currently exist, and "shamanic" techniques and methods are taught throughout North America. The terms no longer refer to any specific cultural phenomenon, but rather to a set of behavioral techniques or methods that are free of any specific cultural context. As such, the phenomenon currently described by the terms *shamanism* and *shaman* are not culturally or ethnically bound.

In regards to the themes noted at the beginning of the book, the phenomenon apparently continues to occur throughout North America, and that the occurrence of the phenomenon of shamanism is not limited to traditional American Indian or First Nation peoples. Likewise, the prevalence of the phenomenon across time indicates that it occurs in many individuals, and that aspects of shamanic techniques and behaviors may work on the human biological, cognitive, or neural level. Finally, as I have argued in this book as a result of examining the comprehensive bibliography, the form the phenomenon has taken has changed over time. Its manifestation today is very different then it was when it was first described in the mid-nineteenth century, and to some extent today's "shamanic" activities have no relationship to those of a hundred years ago.

The comprehensive bibliography and the history of the uses of these terms agrues that: the term *shamanism* defines a phenomenon that occurs among only a select few individuals (the "shamans") at any given spatiotemporal moment, and that operational definitions of the phenomenon must be limited to those specific spatiotemporal moments; and (2) only a nominal definition is possible when discussing the phenomenon diachronically or cross-culturally, and this nominal definition is contingent upon etically understood notions of the folk epistemology and folk ontology of the cultures compared. Though the phenomenon may have been first identified, studied, and defined based on etically observed behavioral traits of certain American Indian and First Nation peoples, the data indicate that it is not the result of specific cultural settings *per se*, as the early historical particularists believed (Boas 1896/1948, 1911; Kroeber 1923; Radin 1914), or as some contemporary radical pro-indigenous scholars maintain (Aldred 2000; Franefort and Hamayon 2002; Kehoe 1981, 2000; York 2001).

Similarly, current definitions of *shamanism* work only in particular spatiotemporal contexts, and that these definitions cannot be used in nominal level investigations. As such, cross-cultural or neurophenomenological discussions of the phenomena are largely circumscribed because the definitions used in these discussions have been operationalized, limiting them to specific spatiotemporal settings; they cannot be used in conjunction with spatiotemporally free cross-cultural and neuro-phenomenological theories.

Peter N. Jones

NOTES

1. Throughout this book, I use the term *shamanism* in a nominal fashion to cover the disparate phenomena that have been associated with it throughout history. I use the term *shaman* to refer to the individual who is the center of this phenomenon.

2. Technically, university-trained anthropologists did not begin to study the phenomenon in North America until the early 20th century. However, I use the term anthropologists to indicate that the early 19th century reports by various explorers, ethnographers, fur trappers, and the like are usually considered anthropological data, and that these early individuals can be considered anthropologists who lacked specific university training.

3. The fact that the law of sympathy has no causal power has been shown to be erroneous, though thus far the causal power of this law has proven to have only minor effects (Dunne and Jahn 1992; Dunne, et al. 1988; Jahn, et al. 2000; Jahn, et al. 1987).

4. In 1871, Tylor borrowed the German word *Kultur*, where it had become well recognized since its first appearance in the 1793 German dictionary of Adelung (1793). The original use in the German language meant the process of cultivation or the degree to which it has been carried (stemming from the Latin *cultura*, which means "growing, cultivation," and *colere*, meaning "cultivate"). Tylor (1871/1924; 1881/1924), however, radically changed this definition to one that meant a state or condition, sometimes described as *extraorganic* or *superorganic*, in which all human societies share.

Peter N. Jones

REFERENCES CITED

Abel, Theodore
   1970 The Foundation of Sociological Theory. New York, NY: Random House.

Abu-Lughod, Lila, ed.
   1991 Writing Against Culture. Santa Fe: SAR Press.

Adelung, J.C.
   1793 Grammatisch-kritisches Worterbuch der hochdeutschen Mundart. Leipzig.

Aldred, Lisa
   2000 Plastic Shamanism and Astroturf Sun Dances. American Indian Quarterly 24(3):329-351.

Anonymous
   193[n.d.] Shawn and His Men Dancers. *In* Redpath Chautauqua Collection, University of Iowa Libraries, Special Collections Department. Iowa City, IA.

Axtell, Horace, and Margo Aragon
    1997 A Little Bit of Wisdom: Conversations with a Nez Perce
    Elder. Norman, Ok: University of Oklahoma Press.

Barnard, Alan
    2000 History and Theory in Anthropology. Cambridge, UK:
    Cambridge University Press.

Bednarik, Robert G., J. D. Lewis-Williams, and Thomas A. Dowson
    1990 On Neuropsychology and Shamanism in Rock Art. Current
    anthropology 31(1):77-84.

Benedict, Ruth Fulton
    1922 The Vision in Plains Culture. American Anthropologist
    24(1):1-23.

Boas, F.
    1896/1948 Race, Language and Culture. New York, NY:
    Macmillan.
    1911 The Mind of Primitive Man. New York, NY: Macmillan.
    1930 Anthropology. *In* Encyclopedia of the Social Sciences.
    E.R.A. Seligman and A. Johnson, eds. Pp. 73-110, Vol. 2. New
    York, NY: Macmillan.
    1933 Relations Between North-West America and North-East
    Asia. *In* The American Aborigines: Their Origin and Antiquity. D.
    Jenness, ed. Pp. 357-370. New York, NY: Cooper Square
    Publishers.

Bock, Philip K.
    1988 Rethinking Psychological Anthropology: Continuity and
    Change in the Study of Human Action. New York, NY: W.M.
    Freeman.

Bock, Philip K., ed.
    1994 Handbook of Psychological Anthropology. London, UK:
    Greenwood Press.

Boyer, L. Bryce

1962 Remarks on the Personality of Shamans with Special Reference to the Apache of the Mescalero Indian Reservation. *In* The Psychoanalytic Study of Society. W. Muensterberger and S. Axelrad, eds. Pp. 233-254, Vol. 2. New York, NY: International Universities Press.

1964a Comparisons of the Shamans and Pseudoshamans of the Apaches of the Mescalero Indian Reservation: A Rorschach Study. Journal of Projective Techniques and Personality Assessment 28:173-180.

1964b Further Remarks Concerning Shamans and Shamanism. Israel Annals of Psychiatry and Related Disciplines 2:235-257.

Bravo, G., and C. Grob

1989 Shamans, Sacraments, and Psychiatrists. Journal of Psychoactive Drugs 21(1):123-8.

Cardena, E.

1999 "You Are Not Your Body": Commentary on "The Motivations for Self-injury in Psychiatric Inpatients". Psychiatry 62(4):331-333.

Crapanzano, Vincent

1992 The Postmodern Crisis: Discourse, Parody, Memory. *In* Rereading Cultural Anthropology. G.E. Marcus, ed. Pp. 87-102. Durham, NC: Duke University Press.

d'Aquili, Eugene G., and Andrew B. Newberg

1998 The Neuropsychological Basis of Religions, or Why God Won't Go Away. Zygon 33(2):187-201.

1999 The Mystical Mind: Probing the Biology of Religious Experience. Minneapolis, MN: Fortress Press.

2000 The Neuropsychology of Aesthetic, Spiritual, and Mystical States. Zygon 35(1):39-51.

Dixon, Roland B.

1904 Some shamans of Northern California. Journal of American Folklore 17:23-27.

1908 Some Aspects of the American Shaman. Journal of American Folklore 21:1-12.

Dobkin de Rios, M., and M. Winkelman
1989 Shamanism and Altered States of Consciousness: An
Introduction. Journal of Psychoactive Drugs 21(1):1-7.

Downton, J. V.
1989 Individuation and Shamanism. The Journal of Analytical
Psychology 34(1):73-88.

Drury, Nevill
2003 Magic and Witchcraft - From Shamanism to Technopagans.
TLS, the Times Literary Supplement (5251):30.

Du Bois, C.
1944 The People of Alor: A Social-Psychological Study of an
East-Indian Island. Minneapolis, MN: University of Minnesota
Press.

Dunne, B.J., and R.G. Jahn
1992 Experiments in Remote Human/Machine Interaction.
Journal of Scientific Exploration 6(4):311-332.

Dunne, B.J., R.D. Nelson, and R.G. Jahn
1988 Operator-Related Anomalies in a Random Mechanical
Cascade. Journal of Scientific Exploration 2(2):155-179.

Durkheim, Emile
1915 The Elementary Forms of the Religious Life. New York,
NY: The Free Press.

Eliade, Mircea
1951/1974 Shamanism: Archaic Techniques of Ecstasy.
Cambridge, MA: Princeton University Press.
1953/1968 Patterns in Comparative Religion. New York, NY:
New American Library.

Erdmann, E.
2003 Borrowed Feathers. Shaman Medicine on the Positive List.
Chirurg 74(3):M83-4.

Flaherty, Gloria
　　1992　Shamanism and the Eighteenth Century. Princeton, NJ:
　　Princeton University Press.

Flaherty, Gloria, and Karl J. Fink
　　1995　Shamanism and the Eighteenth Century. Comparative
　　Literature Studies 32(3):2.

Fletcher, Alice
　　1891　Ethnologic Gleanings Among the Nez Perces. *In* National
　　Anthropological Archives, Smithsonian Institution, Fletcher-La
　　Flesche Papers. Washington, D.C.

Franefort, Henri-Paul, and Roberte Hamayon
　　2002　The Concept of Shamanism, Uses and Abuses. La
　　Recherche 33(357):4.

Frazer, James G.
　　1890/1993　The Golden Bough. New York, NY: Gramercy
　　Books.

Gale, Albert, and Martha Gale
　　19[n.d.]　Songs and Stories of the Red Man. *In* Redpath
　　Chautauqua Collection. Iowa City, IA.

Gebhardt, Eike
　　2002　A Critique of Methodology. *In* The Essential Frankfurt
　　School Reader. A. Arato and E. Gebhardt, eds. Pp. 371-406. New
　　York, NY: Continuum.

Gellner, Ernest
　　1992　Postmodernism, Reason and Religion. London, UK:
　　Routledge.

Gesner, Alonzo
1884 Report of Warm Springs Agent. *In* Annual Report of the Commissioner of Indian Affairs to the Secretary of the Interior for the Year 1884. Pp. 150-153. Washington, DC: Government Printing Office.
1885 Report of Warm Springs Agent. *In* Annual Report of the Commissioner of Indian Affairs to the Secretary of the Interior for the Year 1885. Pp. 171-176. Washington, DC: Government Printing Office.

Gibson, E.M.
1872 No. 67: Annual Report of E.M. Gibson, Neah Bay Agency, Washington Territory. *In* Annual Report of the Commissioner of Indian Affairs to the Secretary of the Interior for the Year 1872. Pp. 350-352. Washington, DC: Government Printing Office.

Greer, M., and J. Greer
2004 Reply to Kehoe: Rock Art and Shamanism. Plains Anthropologist 49(189):81-84.

Grossinger, Richard
1990 Planet Medicine: From Stone Age Shamanism to Post-Industrial Healing. Berkeley, CA: North Atlantic Books.

Gunn, Sisvan W. A.
1966 Totemic Medicine and Shamanism Among the Northwest American Indians. Journal of the American Medical Association 196:700-706.

Harner, Michael
1990 The Way of the Shaman. New York, NY: HarperCollins.
2002 Shamanic Healing: We Are Not Alone, Vol. 2002: The Foundation for Shamanic Studies.

Harris, Llewellyn
1879 Miraculous Healing Among the Zunis. Juvenile Instructor 14:173-176.

Harris, Marvin
1968 The Rise of Anthropological Theory. New York, NY: Thomas Y. Crowell.

Hedges, Kenneth
1983 The Shamanic Origins of Rock Art. *In* Ancient Images on Stone: Rock Art of the Californias. J.A.v. Tilburg, ed. Pp. 46-61. Los Angeles, CA: Rock Art Archive.

Hoffman, Walter James
1888 Pictography and Shamanistic Rites of the Ojibwa. American Anthropologist 1:209-229.

Honigmann, John J.
1949 Parallels in the Development of Shamanism Among Northern and Southern Athapaskans. American Anthropologist 51:512-514.

Hoppal, M.
1989 Changing Image of the Eurasian Shamans. *In* Shamanism: Past and Present. M. Hoppal and O.J. Von Sadovszky, eds. Pp. 75-90, Vol. 1. Budapest, Hungary: International Society for Trans-Oceanic Research.

Hsu, Francis L.K., ed.
1972 Psychological Anthropology. Cambridge, MA: Schenkman Publishing Company.

Hultkrantz, Ake
1997 Shamanic Healing and Ritual Drama. New York, NY: Crossroad Publishing Company.

Ingerman, S.
2003 Sandra Ingerman, MA. Medicine for the earth, medicine for people. Interview by Bonnie Horrigan. Alternative Therapeutic Health Medicine 9(6):76-84.

Jahn, R.G., et al.
2000 Mind/Machine Interaction Consortium: PortREG Replication Experiments. Journal of Scientific Exploration 14(4):499-555.

Jahn, R.G., B.J. Dunne, and R.D. Nelson
  1987 Engineering Anomalies Research. Journal of Scientific Exploration 1(1):21-50.

Jenkins, P.
  2004 Dream Catchers: How Mainstream America Discovered Native Spirituality. Oxford, UK: Oxford University Press.

Johnson, F.
  1943 Notes on Micmac Shamanism. Primitive Man 16:53-80.

Jones, Peter N.
  2004 Shamanism in North America: A Critical Bibliography of the Phenomenon. Boulder, Co: Bauu Institute.
  2006 Shamanism: An Inquiry into the History of the Scholarly Use of the Term in English-Speaking North America. Anthropology of Consciousness 17(2):4-32.

Kehoe, Alice B.
  1981 Revisionist Anthropology: Aboriginal North America. Current Anthropology 22(5):503-517.
  2000 Shamans and Religion: An Anthropological Exploration in Critical Thinking. Prospect Heights, IL: Waveland Press.

Kelly, Isabel Truesdell
  1936 Chemehuevi Shamanism. *In* Essays in Anthropology Presented to Alfred Louis Kroeber. Pp. 129-142. Berkeley, CA.
  1939 Southern Paiute Shamanism. Berkeley, CA: University of California Press.

King, A. G.
  1960 Shamanism. Obstetrics and Gynecology 16:129-132.

Krippner, S.
  2000 The Epistemology and Technologies of Shamanic States of Consciousness. Journal of Consciousness Studies 7(11-12):93-118.

Krippner, S., and A. Combs
2002a The Neurophenomenology of Shamanism. Re-vision 24(Part 3):46-48.
2002b The Neurophenomenology of Shamanism: An Essay Review. Journal of Consciousness Studies 9(3):77-82.
2002 The Neurophenomenology of Shamanism. *In* Journal of Consciousness Studies, Vol. 9(3):77-82.

Kroeber, A.L.
1923 Anthropology. New York, NY: Harcourt, Brace and Company.
1939 Cultural and Natural Areas of Native North America. University of California Publications in American Archaeology and Ethnology 31(4):211-256.

La Flesche, Francis
1890 The Omaha Buffalo Medicine-Men. Journal of American Folklore 3:215-221.

Lane, James D., et al.
1998 Binaural Auditory Beats Affect Vigilance Performance and Mood. Physiology and Behavior 63(2):249-252.

Langdon, E. Jean
1989 Shamanism as the History of Anthropology. *In* Shamanism: Past and Present. M. Hoppal and O.J. Von Sadovszky, eds. Pp. 53-68, Vol. 1. Budapest, Hungary: International Society for Trans-Oceanic Research.

Layton, Robert
2000 Shamanism, Totemism and Rock Art: Les Chamanes de la Prâehistoire in the Context of Rock Art Research. Cambridge Archaeological Journal 10(1):18.

Leahey, Thomas Hardy
1987 A History of Psychology: Main Currents in Psychological Thought. Englewood Cliffs, NJ: Prentice-Hall.

Leighton, Alexander H. and Dorothea C. Leighton
  1941 Elements of Psychotherapy in Navaho Religion. Psychiatry 4: 515-523.

Lewis-Williams, J. D.
  2003a Debate - Putting the Record Straight: Rock Art and Shamanism. Antiquity 77(295):5.

Lewis, I.M.
  1989 Ecstatic Religion: A Study of Shamanism and Spirit Possession. New York, NY: Routledge.

Malinowski, Bronislaw
  1927 The Father in Primitive Psychology. New York, NY: W.W. Norton & Company.
  1944 A Scientific Theory of Culture and Other Essays. Chapel Hill, NC: University of North Carolina Press.
  1948 Magic, Science and Religion and Other Essays. Glencoe, IL: The Free Press.

Mandelbaum, David G., ed.
  1949 Selected Writings of Edward Sapir in Language, Culture and Personality. Berkeley, CA: University of California Press.

Matthews, Washington
  1888 The Prayer of a Navajo Shaman. American Anthropologist 1:149-170.
  1894 Songs of Sequence of the Navajos. Journal of American Folklore 7:185-194.

Maddox, A.
  1904 The Medicine Man: Ancient Religion of Shamanism Flourishing Today. New York Times, July 17.

Maurer, R. L., Sr., et al.
  1997 Phenomenological Experience in Response to Monotonous Drumming and Hypnotizability. American Journal of Clinical Hypnosis 40(2):130-145.

Mauss, Marcel
    1902/1999 A General Theory of Magic. London, UK:
    Routledge.

McClenon, James
    1997 Shamanic Healing, Human Evolution, and the Origin of
    Religion. Journal for the Scientific Study of Religion 36(3):345-
    354.

McMurry, Robert N.
    19[n.d.] Popular Lectures on Applied Psychology,
    Streamlining Your Personality, Maintaining Mental Health,
    Vocational Guidance, Safety. *In* Redpath Chautauqua Collection,
    University of Iowa Libraries, Special Collections Department. Iowa
    City, IA.

Mikhailovskii, V.M.
    1895a Shamanism in Siberia and European Russia-(Continued).
    Journal of the Anthropological Institute of Great Britain and Ireland
    24:124-158.
    1895b Shamanism in Siberia and European Russia, Being the
    Second Part of "Shamanstvo". Journal of the Anthropological
    Institute of Great Britain and Ireland 24:62-100.

Money, M.
    2000 Shamanism and Complementary Therapy. Complementary
    Therapies in Nursing and Midwifery 6(4):207-212.

Money, Mike
    2001 Shamanism as a Healing Paradigm for Complementary
    Therapy. Complementary Therapies in Nursing & Midwifery
    7:126-131.

Morgan, Lewis Henry
    1877/1985 Ancient Society. Tucson, AZ: University of Arizona
    Press.

Murphy, Jane-M
    1964 Psychotherapeutic Aspects of Shamanism on St. Lawrence Island, Alaska. *In* Magic, Faith and Healing Studies in Primitive Psychiatry Today. A. Kiev, ed. Pp. 53-83. New York, NY: Free Press.

Narby, Jeremy, and Francis Huxley, eds.
    2001 Shamans Through Time: 500 Years on the Path to Knowledge. New York, NY: Jeremy P. Thacher/Putnam.

Neher, Andrew
    1962 A Physiological Explanation of Unusual Behavior in Ceremonies Involving Drums. Human Biology 34:151-160.

Noll, Richard
    1985 Mental Imagery Cultivation as a Cultural Phenomenon: The Role of Visions in Shamanism. Current Anthropology 26(4):443-461.

Olson, Ronald L.
    1961 Tlingit Shamanism and Sorcery. [s.n.]: Kroeber Anthropological Society.

Opler, Morris E.
    1936 Some points of Comparison and Contrast Between the Treatment of Functional Disorders by Apache Shamans and Modern Psychiatric Practice. Baltimore, MD: American Psychiatric Association.
    1946 The Creative Role of Shamanism in Mescalero Apache Mythology. Journal of American Folklore 59:268-281.

Orta, Andrew
    2004 The Promise of Particularism and the Theology of Culture: Limits and Lessons of "Neo-Boasianism". American Anthropologist 106(3):473-487.

Ortner, Sherry B.
    1984 Theory in Anthropology since the Sixties. Comparative Studies in Society and History 26:126-166.

Park, W. Z.
    1934 Paviotso Shamanism. American Anthropologist 36:98-113.
    1938 Shamanism in Western North America: A Study in Cultural
    Relationships. Evanston, IL: Northwestern University Press.

Peters, Karl E.
    2001 Neurotheology and Evolutionary Theology: Reflections on
    The Mystical Mind. Zygon 36(3):493-500.

Polimeni, J., and J. P. Reiss
    2002 How Shamanism and Group Selection May Reveal the
    Origins of Schizophrenia. Medical Hypotheses 58(3):5.

Pond, G. H.
    1854 Power and Influence of Dakota Medicine-Men. *In*
    Information Respecting the History, Condition, and Prospects of the
    Indian Tribes of the United States. H.R. Schoolcraft, ed. Pp. 641-
    651, Vol. 4. Philadelphia, PA: [s.n.].

Porterfield, Amanda
    1984 Native American Shamanism and the American Mind-Cure
    Movement: A Comparative Study of Religious Healing. Horizons
    11(2):276-289.

Posinsky, Sollie H.
    1965 Yurok shamanism. Psychiatric Quarterly 39:227-243.

Powell, J.W.
    1899 Eighteenth Annual Report of the Bureau of American
    Ethnology to the Secretary of the Smithsonian Institution, 1896-97,
    Part 1. Washington, DC: Government Printing Office.

Rabinowitz, E.
    2003 Care That Bridges Worlds. Healthplan 44(4):14-8.

Radin, Paul
 1911 The Ritual and Significance of the Winnebago Medicine
 Dance. The Journal of American Folklore 24(92):149-209.
 1914 Religion of the North American Indians. The Journal of
 American Folklore 27(106):335-373.

Ray, Verne F.
 1936 The Kolaskin Cult: A Prophet Movement of 1870 in
 Northeastern Washington. American Anthropologist 38(1):67-75.

Riba, Jordi, et al.
 2001 Subjective Effects and Tolerability of the South American
 Psychoactive Beverage Ayahuasca in Healthy Volunteers.
 Psychopharmacology 154:85-95.

Riches, David
 1994 Shamanism: The Key to Religion. Man 29(2):381-405.

Ripinsky-Naxon, Michael
 1993 The Nature of Shamanism. New York, NY: State University
 of New York Press.

Samuel, Geoffrey
 1996 Nature Religion Today: Western Paganism, Shamanism and
 Esotericism in the 1990s Conference at the Lake District Campus of
 Lancaster University, 9th to 13th April 1996. Religion 26(4):4.

Sapir, Edward
 1949 Culture, Language and Personality. Berkeley, CA:
 University of California Press.

Schneider, G. W., and M. J. DeHaven
 2003 Revisiting the Navajo Way: Lessons for Contemporary
 Healing. Perspectives in Biological Medicine 46(3):413-427.

Schultes, Richard Evans
 1998 Antiquity of the Use of New World Hallucinogens. The
 Heffter Review of Psychedelic Research 1:1-7.

Shamdasani, Sonu
    2003 Jung and the Making of Modern Psychology: The Dream of a Science. Cambridge, UK: Cambridge University Press.

Shimoji, A., and T. Miyakawa
    2000 Culture-bound Syndrome and a Culturally Sensitive Approach: From a Viewpoint of Medical Anthropology. Psychiatry and Clinical Neuroscience 54(4):461-466.

Smith, Harlan Ingersoll
    1896 Certain Shamanistic Ceremonies Among the Ojibwas. American Antiquarian and Oriental Journal 18:282-284.

Smith, Marian W.
    1954 Shamanism in the Shaker Religion of Northwest America. Man 54:119-122.

Speck, Frank G.
    1919 Penobscot shamanism. Menasha, WI: American Anthropological Association.

Taylor, Eugene, and Janet Pledilato
    2002 Shamanism and the American Psychotherapeutic Counter-Culture. Journal of Ritual Studies 16(2):129-140.

Titiev, Mischa
    1956 Shamans, Witches and Chiefs Among the Hopi. Tomorrow 4(3):51-56.

Townsend, Joan B.
    2004 Individualist Religious Movements: Core and Neo-shamanism. Anthropology of Consciousness 15(1):1-9.

Tylor, E.B.
    1871/1924 Primitive Culture. New York, NY: Brentano's.
    1881/1924 Anthropology: An Introduction to the Study of Man and Civilization. New York, NY: D. Appleton.
    1924/1871 Primitive Culture. New York, NY: Brentano's.
    1924/1881 Anthropology: An Introduction to the Study of Man and Civilization. New York, NY: D. Appleton.

Vogel, K.
    2003 Female Shamanism, Goddess Cultures, and Psychedelics.
    Re-vision 25(Part 3):18-29.

von Keitz, E.
    1999 Medicine Men. Journal of Christian Nursing 16(4):26-27.

von Stuckard, K.
    2002 Reenchanting Nature: Modern Western Shamanism and
    Nineteenth-century Thought. Journal of the American Academy of
    Religion 70(4):771-800.

Wallerstein, Immanuel
    2003 Anthropology, Sociology, and Other Dubious Disciplines.
    Current Anthropology 44(4):453-465.

Waters, Frank
    1950 Masked Gods: Navaho and Pueblo Ceremonialism.
    Albuquerque, NM: University of New Mexico Press.

Wellmann, Klaus F.
    1981 Rock Art, Shamans, Phosphenes and Hallucinogens in North
    America. Bollettino del Centro Camuno di Studii Preistorici 18:89-
    103.

Wiercinski, Andrzej
    1989 On the Origin of Shamanism. *In* Shamanism: Past and
    Present. M. Hoppal and O.J. Von Sadovszky, eds. Pp. 19-25, Vol. 1.
    Budapest, Hungary: International Society for Trans-Oceanic
    Research.

Winkelman, Michael
    1982 Magic: A Theoretical Reassessment. Current Anthropology
    23(1):37-66.
    1984a A Cross-Cultural Study of Magico-Religious Practitioners.
    In Proceedings of the International Conference on Shamanism. R.-
    I. Heinze, ed, Vol. 27-38. Berkeley, CA: Independent Scholars of
    Asia.

1984b A Cross-Cultural Study of Magico-Religious Practitioners. Dissertation, University of California, Irvine.

1989 A cross-cultural study of shamanistic healers. Journal of Psychoactive Drugs 21(1):17-24.

1991 Physiological and Therapeutic Aspects of Shamanistic Healing. Subtle Energies 1(2):1-18.

1995 Psychointegrator Plants: Their Roles in Human Culture, Consciousness and Health. Yearbook of Cross-Cultural Medicine and Psychotherapy:9-53.

1996 Shamanism and Consciousness: Metaphorical, Political and Neurophenomenological Perspectives. Transcultural psychiatric research review 33(1):69.

2000 Shamanism: The Neural Ecology of Consciousness and Healing. Westport, CT: Bergin & Garvey.

2002 Shamanism as Neurotheology and Evolutionary Psychology. American Behavioral Scientist 45(12):1873-1885.

2003 Complementary therapy for addiction: "drumming out drugs". Am J Public Health 93(4):647-51.

2004 Shamanism as the Original Neurotheology. Zygon 39(1):193-217.

Wissler, Clark
1914 Material Cultures of the North American Indians. American Anthropologist 16(3):447-505.

1916 General Discussion of Shamanistic and Dancing Societies. Pp. 853-876. New York, NY: American Museum of Natural History.

Wundt, W.M.
1896 Lectures on Human and Animal Psychology. New York: Macmillan.

York, Michael
2001 New Age Commodification and Appropriation of Spirituality. Journal of Contemporary Religion 16(3):361-372.

Zatzick, D. F., and F. A. Johnson
1997 Alternative Psychotherapeutic Practice Among Middle Class Americans: II: Some Conceptual and Practical Comparisons. Culture, Medicine, and Psychiatry 21(2):213-246.

Peter N. Jones

THE
COMPREHENSIVE
BIBLIOGRAPHY

In compiling the bibliography, numerous libraries, archives, and data-bases were searched. In all, 780 individual documents were located that specifically referenced the words *shamanism* or *shaman* in the context of North America (Jones 2006). Of these, 454 (58.2 percent) of them dealt explicitly with American Indians or First Nation peoples when using the terms. The earliest document found that used the actual term *shaman* was that of G. H. Pond (1854) and his discussion of Dakota medicine men. The first scholars to use the terms were largely early anthropologists and government officials in their discussion of American Indian and First Nation peoples during the late 1800s (see, for example, Fletcher 1891; Gesner 1884; Gesner 1885; Gibson 1872; Harris 1879; La Flesche 1890; Matthews 1888; Powell 1899). Historically, the terms were used exclusively to refer to individuals who were associated with American Indian and First Nation peoples, a process which continued through the remainder of the 19th century and into the early 20th century. By the 1930s, however, the use of the terms expanded outside of anthropology and began to be used by self-help oriented lay psychologists, theater performers, and others (see, for example, Anonymous 193[n.d.]; Gale and Gale 19[n.d.]; McMurry

47

19[n.d.]). Among scholars the terms were still exclusively used within anthropological and religious studies literature until the 1960s, when they first also began to be used within the psychological literature (e.g., Boyer 1962; Boyer 1964a; Boyer 1964b; Murphy 1964). This expansion of use into the psychological literature can largely be credited to humanistic and transpersonal psychologists who were becoming acquainted with the anthropological literature at this time (Leahey 1987; Taylor and Pledilato 2002). Presently, the terms are used in almost every academic field, ranging from nursing and biomedical literature (Rabinowitz 2003; Schneider and DeHaven 2003; Winkelman 2003) to neuroscience (Krippner and Combs 2002b; Polimeni and Reiss 2002; Shimoji and Miyakawa 2000), to religious studies and social science literature (to list a few, see Drury 2003; Taylor and Pledilato 2002; Townsend 2004; Vogel 2003; von Stuckard 2002; York 2001). However, all of these fields still largely rely on anthropological accounts and definitions of the terms, demonstrating the central role anthropology has had in the understanding and uses of these terms in North America.

As the comprehensive bibliography included in this book also indicates, from the first documented uses of the terms *shamanism* and *shaman* within the literature of North America in 1854 until the present, the terms have grown in range of use and broadness of definition. For example, the terms were originally employed to refer to specific individuals within particular American Indian and First Nation groups, but by the 1930s, the terms were beginning to be used to refer to techniques, dance styles, and various paraphernalia. That is, by the 1930s, the terms were being employed outside their original area of usage by anthropologists who used it to refer to individuals within particular sociocultural settings, and instead came to be used in reference not to individuals, but to techniques, characteristics, and processes that were separate from the individual or sociocultural setting they were originally defined by. Part of this change in the understanding of these terms may be due to the simultaneous rise of behaviorism, functionalism, and structuralism that was sweeping through not only psychology, but also anthropology and sociology at the time (Abel 1970; Barnard 2000; Harris 1968; Leahey 1987). In essence, scholars turned away from the individual and their sociocultural role, as well as discussions of that individual's experiences, and focused more on behavioral techniques, structures, and functions over the course of the term's useage in North America.

In 1951 Eliade (1953/1968) published the first cross-cultural examination of shamanism, not only basing his study on the comparative method but also ushering in the official study of shamanism as a scientific field of study within the domain of religious studies. This broad comparative study led Eliade to conclude that shamanism "is precisely one of the archaic techniques of ecstasy — at once mysticism, magic, and 'religion' in the broadest sense of the term" (1951/1974:xix). An unforeseen result of developing the study of shamanism into its own field was that the uses of the terms became divorced from their own specific sociocultural ontogenies, and were instead assumed to define a universal phenomenon that all individuals and all cultures theoretically possessed. This allowed scholars and practitioners to not only radically change the understanding of *shamanism* and *shaman* to fit their specific needs based on etic, reductionistic, and behavioral categorical constructions, but it also required the phenomenon to be reduced to a simple set of operationally circumscribed characteristics in the construction of the term's newly emerging nominal definition. The result is that today the terms are widely used to designate any individual, irrespective of their sociocultural setting, who practices certain techniques or characteristics often associated with some form of "healing" (a term that has its own theoretical and philosophical problems). Thus, the original role of the shaman within cultures of North America has radically shifted from an individual who can manipulate the weather; who is both considered malevolent and benevolent; who is both feared and respected within their culture; who must experience a radical form of a calling; who can manipulate their appearance (that is, shape-shift); who at any moment may lose their special abilities if particular physical and metaphysical precautions are not taken; who helps with the subsistence regime of the culture; and who partakes in many other activities to the present understanding consisting of techniques and characteristics that almost any individual can perform. This radical shift in the nominal definitions of the terms *shamanism* and *shaman* are documented in the comprehensive bibliography included in this book.

Peter N. Jones

SHAMANISM IN NORTH AMERICA:

A COMPREHENSIVE BIBLIOGRAPHY ON THE USE OF THE TERM

Peter N. Jones

Abel, K. M. (1986). Prophets, priests and preachers: Dene Shamans and Christian Missions in the Nineteenth Century. Canadian Historical Association, Historical Papers, 211-224.

Achterberg, J. (1985). Imagery in Healing: Shamanism and Modern Medicine. Boston, MA: Shambala.

Adair, J. (1963). Physicians, Medicine Men and their Navaho Patients. In Galdston, I. (Ed.), Man's Image in Medicine and Anthropology (pp. 237-257). New York, NY: International Universities Press.

Adlam, R. G. (1998). Les Marmottes, Les Femmes et Les Esprits Gardiens: Transformation Spirituelle et Chamanisme. Recherches Amerindiennes au Quebec, 28(3-4), 41-48.

Aggiar, G. (1974). Shamans. Inumarit, 3(2), 1-3.

Aldred, L. (2000). Plastic Shamanism and Astroturf Sun Dances. American Indian Quarterly, 24(3), 329-351.

Almstedt, R. F. (1977). Diegueno Curing Practices. San Diego, CA: San Diego Museum of Man.

Alsup, R., & Krippner, S. (1996). The Mythology of Evil Among North American Indian Yuroks and Its Implications for Western Spirituality. Anthropology of Consciousness, 7(3), 15-29.

Anderson, D. G. (1999). Circumpolar Animism and Shamanism; edited by Takako Yamada and Takashi Irimoto. Arctic, 52(1), 100.

Anderson, W. L., & Bisson, R. (1984). Palisot de Beauvois and Cherokee Snakebite Remedies. Journal of Cherokee Studies, 9, 4-9.

Andrews, L. V. (1986). Star Woman We Are Made From Stars and to the Stars We Must Return. New York, NY: Warner Books.

Andrews, L. V. (1987). Crystal Woman the Sisters of the Dreamtime. New York, NY: Warner Books.

Andrews, L. V., & Momaday, N. S. (1984). Flight of the Seventh Moon: The Teaching of the Shields. San Francisco, CA: Harper & Row.

Andrews, L. V., & Reeves, D. (1981). Medicine Woman. San Francisco, CA: Harper & Row.

Angulo, J. d. (1975). Portrait of a Young Shaman. Alcheringa, 1(1), 9-12.

Angulo, J. d. (1975). Shaman Songs. Alcheringa, 1(1), 25-26.

Angulo, J. d., & Scanzoni, M. (1983). Indianer im Overall. Mit einer biographischen Skizze von Gui de Angulo, aus dem Amerikanischen von M. Scanzoni. Munchen, Germany: Trickster Verlag.

Anhauser, M. (2003). Pharmacists Seek the Solution of a Shaman. Drug Discov Today, 8(19), 868-9.

Anisimov, A.F. (1963). The Shaman's Tent of the Evenks and the Origin of the Shamanistic Rite. In Henry Micael (Ed.), Siberian Shamanism (pp. 84-123). Toronto: Toronto University Press.

Anonymous. (193[n.d.]). Shawn and His Men Dancers.Unpublished manuscript, Iowa City, IA.

Anonymous. (1996). Nation Iroquoise: Abrege des vies et moeurs et autres particularitez de la Nation Irokoise laquelle est divisee en Cinq villages. Scavoir Agnez Onnwyt Nontague Goyogan et Sonnontan. Recherches Amerindiennes au Quebec, 26(2), 31-36.

Armer, L. A. (1953). The Crawler, Navaho Healer. Masterkey, 27, 5-10.

Arthos Jr, J. (2001). The Shaman-Trickster's Art of Misdirection: The Rhetoric of Farrakhan and the Million Men. The quarterly journal of speech, 87(1), 20.

Asatchaq, T., & Lowenstein, T. (1992). The Things that Were Said of them: Shaman Stories and Oral Histories of the Tikigaq People. Berkeley, CA: University of California Press.

Asher, B. (1994). A Shaman-Killing Case on Puget Sound, 1873-1874: American Law and Salish Culture. Pacific Northwest Quarterly, 86(1), 17-23.

Atkinson, J. M. (1992). Shamanisms Today. Annual Review of Anthropology, 21, 307-330.

Atwood, M. D. (1991). Spirit Healing: Native American Magic and Medicine. New York, NY: Sterling Publishing.

Bahr, D. M. (1973). Psychiatry and Indian Curing. Indian Programs, 2(4), 1-9.

Bahr, D. M. (1977). Breath in Shamanic Curing. In Blackburn, T. C. (Ed.), Flowers of the Wind (pp. 29-40). Socorro, NM: Ballena Press.

Bahr, D. M. (1983). Pima and Papago Medicine and Philosophy. In Ortiz, A. (Ed.), Southwest (Vol. 10, pp. 193-200). Washington, DC: Smithsonian Institution.

Bahr, D. M., Gregorio, J., Lopez, D. I., & Alvarez, A. (1974). Piman Shamanism and Staying Sickness (Ka:cim Mumkidag). Tucson, AZ: University of Arizona Press.

Baker, J. P. (1992). The Shamanic Dimensions of Childbirth. Pre- and Peri-natal Psychology Journal, 7(1), 5-21.

Balikci, A. (1963). Shamanistic Behavior Among the Netsilik Eskimos. Southwestern Journal of Anthropology, 19, 380-396.

Balikci, A. (1967). Shamanistic Behavior Among the Netsilik Eskimos. In Middleton, J. (Ed.), Magic, Witchcraft, and Curing (pp. 191-209). Garden City, NJ: Natural History Press.

Barbeau, C. M. (1958). Medicine Men on the North Pacific Coast. Ottawa, Ontario: National Museum of Canada.

Bard, J. C., Busby, C. L., & Findlay, J. M. (1981). A Cultural Resources Overview of the Carson and Humboldt Sinks, Nevada (No. Cultural Resource Series No. 2). Carson City, NV: Bureau of Land Management.

Barkow, J. H. (1974). Proceedings of the First Congress, Canadian Ethnology Society. Paper presented at the First Congress of the Canadian Ethnology Society, Ottawa, Ontario.

Barrett, S. A. (1917). Pomo Bear Doctors. Berkeley, CA: University of California Press.

Bartels, D. (1985). Shamanism, Christianity, and Marxism: Comparisons and Contrasts Between the Impact of Soviet Teachers on Eskimos, Chukchis, and Koryaks in Northeastern Siberia, and the Impact of an Early Anglican Missionary on Baffin Island Inuit. Canadian Journal of Native Education, 12(3), 1-7.

Basso, K. H. (1989). Southwest: Apache. In Walker, D. E., Jr. (Ed.), Witchcraft and Sorcery among American Native Peoples (pp. 167-190). Moscow, ID: University of Idaho Press.

Beals, R. L. (1978). Sonoran Fantasy or Coming of Age? American Anthropologist, 80, 355-362.

Bean, L. J. (1976). California Indian Shamanism and Folk Curing. Paper Presented at the American Folk Medicine: A Symposium, Berkeley, CA.

Bean, L. J. (1991). The Artistic and the Shamanic Tradition. In Bean, L. J. & Vane, S. B. (Eds.), Ethnology of the Alta California Indians-Volume 2: Postcontact (pp. 963-970). New York, NY: Garland Publishing.

Bean, L. J. (1991). California Indian Shamanism and Folk Curing. In Bean, L. J. & Vane, S. B. (Eds.), Ethnology of the Alta California Indians-Volume 1: Precontact (pp. 725-739). New York, NY: Garland Publishing.

Bean, L. J., & Blackburn, T. C. (1976). Native Californians: A Theoretical Retrospective. Ramona, CA: Ballena Press.

Bean, L. J., & Vane, S. B. (1978). Shamanism: An Introduction. In Berrin, K. (Ed.), Art of the Huichol Indians (pp. 118-128). New York, NY: Harry N. Abrams.

Bean, L. J., & Vane, S. B. (1991). Shamanism: An Introduction. In Bean, L. J. & Vane, S. B. (Eds.), Ethnology of the Alta California Indians-Volume 1: Precontact (pp. 713-723). New York, NY: Garland Publishing.

Bear, L. (1980). The Sacred Language [August 1900]. Lincoln, NE: University of Nebraska Press.

Beaudry, N. (1998). Perches magiques et bossus masques: Magie, jeu ou rituel? Recherches Amerindiennes au Quebec, 28(3-4), 19-26.

Bednarik, R. G., Lewis-Williams, J. D., & Dowson, T. A. (1990). On Neuropsychology and Shamanism in Rock Art. Current Anthropology, 31(1), 77-84.

Begay, D. H., & Maryboy, N. C. (2000). The Whole Universe is My Cathedral: A Contemporary Navajo Spiritual Synthesis. Medical Anthropology Quarterly, 14(4), 498-520.

Beiser, M., & Degroat, E. (1974). Conversations with a Navajo Singer. Psychiatric Annals, 4(9), 9-12.

Bend, C., & Wiger, T. (1988). Birth of a Modern Shaman: A Documented Journey and Guide to Personal Transformation. St. Paul, MN: Llewellyn Publications.

Benedict, R. F. (1922). The Vision in Plains Culture. American Anthropologist, 24(1), 1-23.

Bergman, R. L. (1973). A School for Medicine Men. American Journal of Psychiatry, 130(6), 663-666.

Berry, R. V. S. (1929). The Navajo Shaman and His Sacred Sand-Paintings. Art and Archaeology, 27, 3-17.

Biesele, M.A. (1979). Old K'xau. In J. Halifax (Ed.), Shamanic Voices: A Survey of Visionary Narratives (pp. 54-62). New York: E.P. Dutton.

Blacher, R. S. (1984). The Briefest Encounter: Psychotherapy for Medical and Surgical Patients. General Hospital Psychiatry, 6(3), 226-32.

Black Elk, W., & Lyon, W. S. (1990). Black Elk: The Sacred Ways of a Lakota. San Francisco, CA: Harper & Row.

Black Rogers, M. B. (1989). Dan Rainclown "Keeping our Indian Way". In Clifton, J. A. (Ed.), Being and Becoming Indian: Biographical Studies of North American Frontiers (pp. 226-248). Chicago, IL: The Dorsey Press.

Blackburn, T. C. (1977). Flowers of the Wind: Papers on Ritual, Myth, and Symbolism in California and the Southwest. Socorro, NM: Ballena Press.

Blackburn, T. C. (1977). Introduction. In Blackburn, T. C. (Ed.), Flowers of the Wind (pp. 7-9). Socorro, NM: Ballena Press.

Blanchard, D. S. (1982). Who or What's a Witch? Iroquois Persons of Power. American Indian Quarterly, 6, 218-237.

Blerkom, L. M. V. (1995). Clown Doctors: Shaman Healers of Western Medicine. Medical Anthropology Quarterly, 9(4), 14.

Blodgett, J. (1978). The Coming and Going of the Shaman: Eskimo Shamanism and Art. Winnipeg, Manitoba: Winnipeg Art Gallery.

Blumensohn, J. (1933). The Fast among North American Indians. American Anthropologist, 35(3), 451-469.

Bobrow, R. S. (2003). Paranormal Phenomena in the Medical Literature Sufficient Smoke to Warrant a Search for Fire. Medical Hypotheses, 60(6), 864-8.

Bogoras, W. (1919). O tak nazyvaemom iazykie dukhov (shamanskom). Akademiia Nauk SSSR, Izvestiia, 6(8/11), 489-495.

Bogoras, W. (1958). Shamanic Performance in the Inner Room. In W.A. Lessa and E.Z. Vogt (Eds.), Reader in Comparative Religion (pp. 382-387). New York: Harper and Row.

Borsboom, A. P. (1984). Shaman: Magician or Healer? Tijdschr Ziekenverpl, 37(25), 781-786.

Bourguignon, E. (1989). Trance and Shamanism: What's in a Name? Journal of Psychoactive Drugs, 21(1), 9-15.

Bourke, J. G. (1892). The Medicine Men of the Apache. Washington, DC: U.S. Bureau of American Ethnology.

Bourke, J. G. (1970). The Medicine Men of the Apache. Glorieta, NM: Rio Grande Press.

Bourke, J. G. (1993). Apache Medicine-Men. New York, NY: Dover Publications.

Bouteiller, M. (1950). Don chamanistique et adaptation a la vie. Societe des Americanistes, 39, 1-14.

Boyd, D. (1974). Rolling Thunder. New York, NY: Delta Books.

Boyer, L. B. (1962). Remarks on the Personality of Shamans with Special Reference to the Apache of the Mescalero Indian Reservation. In Muensterberger, W. & Axelrad, S. (Eds.), The Psychoanalytic Study of Society (Vol. 2, pp. 233-254). New York, NY: International Universities Press.

Boyer, L. B. (1964). Comparisons of the Shamans and Pseudoshamans of the Apaches of the Mescalero Indian Reservation: A Rorschach Study. Journal of Projective Techniques and Personality Assessment, 28, 173-180.

Boyer, L. B. (1964). Further Remarks Concerning Shamans and Shamanism. Israel Annals of Psychiatry and Related Disciplines, 2, 235-257.

Boyer, L. B. (1969). Shamans: To Set the Record Straight. American Anthropologist, 71, 307-309.

Boyer, L. B., & Boyer, R. M. (1977). Understanding the Individual Through Folklore. Contemporary Psychoanalysis, 13(1), 30-51.

Boyer, L. B., Boyer, R. M., & DeVos, G. A. (1982). An Apache Woman's Account of Her Recent Acquisition of the Shamanistic Status. Journal of Psychoanalytic Anthropology, 5(3), 299-331.

Bramly, S. (1974). Terre Wakan univers sacre des Indiens d'Amerique du Nord. Paris, France: Editions Robert Laffont.

Brasser, T. J. (1973). Wolf Collar: The Shaman as Artist. 70-73.

Braun, P. (1973). Les hommes du grand nord. Paris, France: J.C. Lattes/Edition Speciale.

Bravo, G., & Grob, C. (1989). Shamans, Sacraments, and Psychiatrists. Journal of Psychoactive Drugs, 21(1), 123-8.

Bray Morris, J. (2002). Blood, Sweat, and Tears. Journal of the American Board of Family Practice, 15(4), 332-3.

Brebeuf, J. d., & LeMercier, F. J. (1981). Missionaries Meet the Medicine Men in Huronia (1635, 1637). In Axtell, J. (Ed.), The Indian Peoples of Eastern America (pp. 189-192). New York, NY: Oxford University Press.

Brill, J. (1992). Shamanism--Toddler Stage. Voices; The Art and Science of Psychotherapy, 28(4), 57.

Brotherston, G. (1979). The Seance of an Eskimo Shaman in Alaska. In Brotherston, G. (Ed.), Image of the New World: The American Continent Portrayed in Native Texts (pp. 94-95). London, UK: Thames and Hudson.

Brotherston, G. (Ed.). (1979). The Journey of the Soul After Death CT: Image of the New World: The American Continent Portrayed in Native Texts. London, UK: Thames and Hudson.

Brotherston, G. (1984). "Far as the Solar Walk": The Path of the North American Shaman. Indiana, 9, 15-29.

Brown, J. A. (1997). The Archaeology of Ancient Religion in the Eastern Woodlands. Annual Review of Anthropology, 26, 465-485.

Buchbinder, G. (1988). The Tiger and the Shaman: Towards an Understanding of Why We Tolerate a Major Cause for Death and Disability. Hawaii Medical Journal, 47(9), 441-443.

Butt, A., S. Wavell, & N. Epton. (1966). Trances. London: George Allen and Unwin.

Calestro, Kenneth. (1972). Psychotherapy, Faith Healing, and Suggestion. International Journal of Psychiatry, 10(2), 83-113.

Cardena, E. (1999). "You Are Not Your Body": Commentary on "The Motivations for Self-injury in Psychiatric Inpatients". Psychiatry, 62(4), 331-333.

Carter, W. H. (1973). Medical Practices and Burial Customs of the North American Indians. London, Ontario: Namind Printers and Publishers.

Castaneda, C. (1968). The Teachings of Don Juan a Yaqui Way of Knowledge. Berkeley, CA: University of California Press.

Castaneda, C. (1971). A Separate Reality: Further Conversations with Don Juan. New York, NY: Simon and Schuster.

Castaneda, C. (1974). Tales of Power. New York, NY: Simon and Schuster.

Catches, P., & Catches, P. V. (1999). Sacred Fireplace (Oceti Wakan):Life and Teachings of a Lakota Medicine Man. Santa Fe, NM: Clear Light Publishers.

Chadwick, N. K. (1936). Shamanism Among the Tatars of Central Asia. Journal of the Royal Anthropological Institute of Great Britian and Ireland, 66, 75-112.

Chadwick, N. K. (1936). The Spiritual Ideas and Experiences of the Tatars of Central Asia. The Journal of the Royal Anthropological Institute of Great Britain and Ireland, 66, 291-329.

Chalfant, W. A. (1931). Medicine Men of the Eastern Mono. Masterkey, 5, 50-54.

Chamberlain, A. F. (1901). Kootenay "Medicine Man". Journal of American Folklore, 14, 95-99.

Charet, F. X. (1998). Jung and Shamanism in Dialogue: Retrieving the Soul/Retrieving the Sacred. Nova Religio, 2(2), 327-328.

Chief, W. (1995). Sacred Bundles. In Hirschfelder, A. (Ed.), Native Heritage: Personal Accounts by American Indians, 1790 to the Present (pp. 199). New York, NY: Macmillan.

Ciprian-Ollivier, J., & Cetkovich-Bakmas, M. G. (1997). Altered Consciousness States and Endogenous Psychoses: A Common Molecular Pathway? Schizophrenia Research, 28(2-3), 257-65.

Clapp, R. A., & Crook, C. (2002). Drowning in the Magic Well: Shaman Pharmaceuticals and the Elusive Value of Traditional Knowledge. Journal of Environment & Development, 11(1), 79-102.

Cloud, J. (2001). Recreational Pharmaceuticals. Time, 157(2), 100.

Cloutier, D. (1989). Spirit, Spirit: Shaman Songs: Versions. Providence, RI: Copper Beech Press.

Clune, Francis J. (1973). A Comment on Voodoo Deaths. American Anthropologist, 75(1), 32.

Collings, P. R. (1973). The Huichol Indians: A Look at a Present-day Drug Culture. Part One. Masterkey, 47, 124-133.

Collings, P. R. (1974). The Huichol Indians: A Look at a Present-day Drug Culture. Part Two. Masterkey, 48, 22-29.

Conklin, B. A. (2002). Shamans Versus Pirates in the Amazonian Treasure Chest. American Anthropologist, 104(4),1050-1061.

Connor, L., Thomas, N., & Humphrey, C. (1996). Shamanism, History and the State. Oceania, 67(2), 2.

Conway, T. (1992). The Conjurer's Lore: Celestial Narratives from Algonkian Shamans. In Williamson, R. A. & Farrer, C. R. (Eds.), Earth & Sky: Visions of the Cosmos in Native American Folklore (pp. 236-259). Albuquerque, NM: University of New Mexico Press.

Corlett, W. T. (1935). The Medicine-Man of the American Indian. Unpublished manuscript.

Corlett, W. T. (1935/1977). The Medicine-Man of the American Indian and His Cultural Background. New York, NY: AMS Press.

Coudray, J. P., Delpretti, M., Luccioni, H., & Scotto, J. C. (1972). The Shaman and Psychiatry. Preliminary Note on the Objectivation of Mental Disease. Evolutionary Psychiatry (Paris), 37(1), 131-9.

Coulam, N. J., & Schroedl, A. R. (2004). Late Archaic Totemism in the Greater American Southwest. American Antiquity, 69(1), 41-62.

Cowan, W. (1976). Papers of the Seventh Algonquian Conference, 1975: Papers presented at The Algonquian Conference, 7th, October 31-November 2, 1975, Niagara-on-the-Lake, Ontario. Paper presented at the Seventh Algonquian Conference.

Cox, P. A. (1995). Shaman as Scientist: Indigenous Knowledge Systems in Pharmacological Research and Conservation. Annual Proceedings of the Phytochemical Society of Europe, 37, 1.

Crepeau, R. R. (1997). Le chaman croit-il vraiment a ses manipulations et a leurs fondements intellectuels? Recherches Amerindiennes au Quebec, 27(3-4), 7-17.

Csordas, T. J. (2000). The Navajo Healing Project. Medical Anthropology Quarterly, 14(4), 463-75.

Cutler, J., & Miller, I. (1995). Eno Washington: The Memoirs of a Mississippi Shaman. Race & class, 36(3), 21.

Danielo, E. (1955). The Story of a Medicine Man. Eskimo, 36, 3-6.

Danielo, E. (1955). Une histoire de sorcier. Eskimo, 36, 3-6.

Das, P. (1978). Initiation by a Huichol shaman. In Berrin, K. (Ed.), Art of the Huichol Indians (pp. 129-141). New York, NY: Harry N. Abrams.

Dauenhauer, N. M. (1988). The Droning Shaman Poems. Hains, AK: Black Current Press.

De Laguna, F. (1973). Tlingit Shamans. In Art, N. G. o. (Ed.), The Far North (pp. 227-279). Washington, DC: National Gallery of Art.

De Laguna, F. (1987). Atna and Tlingit Shamanism: Witchcraft on the Northwest Coast. Arctic Anthropology, 24(1), 84-100.

de Mille, R. (1976). Castaneda's Journey: The Power and the Allegory. London, UK: [s.n.].

de Niord, C. (1993). Shaman. The North American Review, 278(2), 10-11.

DeAngulo, J. (1928). La Psychologie Religieuse des Achumawi, IV: Le Chamanisme. Anthropos, 23, 561-582.

Degarrod, L. N. (1998). Female Shamanism and the Mapuche Transformation into Christian Chilean Farmers, Religion, 28, 339-350.

Deutsch, R. D. (1996). A Shaman Looks at Campaign '96. Nieman Reports, 50(3), 41.

Devereux, G. (1957). Dream Learning and Individual Ritual Differences in Mohave Shamanism. American Anthropologist, 59, 1036-1045.

Devereux, G. (1970). Re: Mojave Shamans. Masterkey, 44, 155-156.

Devereux, G. (1976). Mohave Soul Concepts. In Bean, L. J. & Blackburn, T. C. (Eds.), Native Californians: A Theoretical Retrospective (pp. 331-336). Ramona, CA: Ballena Press.

Devereux, P. (2001). Did Ancient Shamanism Leave a Monumental Record on the Land as Well as in Rock Art? BAR International Series (supplementary), 936, 1-8.

Dewdney, S. H. (1970). Ecological Notes on the Ojibway Shaman-Artist. Arts Canada, 27, 17-28.

Dixon, R. B. (1904). Some Shamans of Northern California. Journal of American Folklore, 17, 23-27.

Dixon, R. B. (1908). Some Aspects of the American Shaman. Journal of American Folklore, 21, 1-12.

Dobkin de Rios, M. (2002). What We can Learn from Shamanic Healing: Brief Psychotherapy with Latino Immigrant Clients. American Journal of Public Health, 92(10), 1576-1578.

Dobkin de Rios, M., & Winkelman, M. (1989). Shamanism and Altered States of Consciousness: An Introduction. Journal of Psychoactive Drugs, 21(1), 1-7.

Dodge, E. S. (1985). A Cayuga Bear Society Curing Rite. In Tooker, E. (Ed.), An Iroquois Source Book: Medicine Society Rituals (Vol. 3, pp. 65-71). New York, NY: Garland Publishing.

Dorais, L. J. (1997). Pratiques et sentiments religieux a Quaqtaq: continuite et modernite. Etudes/Inuit/Studies, 21(1-2), 255-267.

Dow, James. (1982). Las Figuras de Papel y el Concepto del Alma entre los Otomies de la Sierra. America Indigena, 42(4), 629-650.

Dow, James (1984). Symbols, Soul, and Magical Healing Among the Otomi Indians. Journal of Latin American Lore, 10(1), 3-21.

Dow, James. (1986). The Shaman's Touch: Otomi Indian Symbolic Healing. Salt Lake City: University of Utah Press.

Downs, J. F. (1961). Washo Religion (Vol. 16). Berkeley, CA: University of California Press.

Downton, J. V. (1989). Individuation and Shamanism. The Journal of Analytical Psychology, 34(1), 73-88.

Drury, C. M. (1976). Nine Years with the Spokane Indians: The Diary, 1838-1848, of Elkanah Walker. Glendale, CA: Arthur H. Clark.

Drury, N. (2003). Magic and Witchcraft - From shamanism to Technopagans. TLS, the Times Literary Supplement, 5251, 30.

Dufrene, P. M. (1990). Utilizing the Arts for Healing from a Native American Perspective: Implications for Creative Arts Therapies. The Canadian Journal of Native Studies, 10(1), 121-131.

Durante, I. I. (1974). L'uomo medicina fra i popoli del Nord-America. Terra Ameriga, 10(31-32), 59-68.

Duvernay, J. (1973). Les voies du chamane. L'Homme, XIII(3), 82-92.

Eaton, E. (1982). The Shaman and the Medicine Wheel. New York, NY: Quest.

Eckstorm, F. H. (1945). Old John Neptune and other Maine Indian Shamans. Portland, ME: [s.n.].

Eide, A. H. (1952). Drums of Diomede: Transformation of the Alaska Eskimo. Hollywood, CA: House-Warven.

Eliade, M. (1974). Shamanism: Archaic Techniques of Ecstasy. Cambridge, MA: Princeton University Press.

Elkin, A.P. (1977). Aboriginal Men of High Degree. New York, NY: St. Martin's Press.

Ellis, C. J. (1994). Miniature Early Paleo-Indian Stone Artifacts from the Parkhill, Ontario Site. North American Archaeologist, 15(3), 253-267.

Emery, J. (1991). Boatbuilder as Shaman. The Woodenboat, 103, 64-71.

Encinales, P. (1979). A Shaman Obeys. In Nabokov, P. (Ed.), Native American Testimony (pp. 77). New York, NY: Harper & Row.

Engebretson, J. (2002). Culture and Complementary Therapies. Complementary Therapies in Nursing and Midwifery, 8(4), 177-84.

Erdmann, E. (2003). Borrowed Feathers. Shaman Medicine on the Positive List. Chirurg, 74(3), M83-4.

Erigaktuk, A. (1976). The Shaman and His Kidnapped Wife. Inuvialiut, 5, 14-16.

Ethridge, R. F. (1978). Tobacco Among the Cherokees. Journal of Cherokee Studies, 3, 76-86.

Etsuko, M. (1991). The Interpretations of Fox Possession: Illness as Metaphor. Culture, Medicine, and Psychiatry, 15(4), 453-77.

Evans, E. (1934). The Other View of Chief Kitsap. Washington Historical Quarterly, 25(4), 299-301.

Farber, C. M. (1984). Afterword: Time in a Box. In Seguin, M. (Ed.), The Tsimshian (pp. 311-322). Vancouver, BC: University of British Columbia Press.

Fenton, W. N. (1940). Masked Medicine Societies of the Iroquois. Washington, DC: Smithsonian Institution.

Fenton, W. N. (1982). Masked Medicine Societies of the Iroquois. In Mathews, Z. P. & Jonaitis, A. (Eds.), Native North American Art History: Selected Readings (pp. 325-365). Palo Alto, CA: Peek Publications.

Fenton, W. N. (1985). Masked Medicine Societies of the Iroquois. In Tooker, E. (Ed.), An Iroquois Source Book: Medicine Society Rituals (Vol. 3, pp. 397-429). New York, NY: Garland Publishing.

Fenton, W. N. (1985). The Seneca Society of Faces. In Tooker, E. (Ed.), An Iroquois Source Book: Medicine Society Rituals (Vol. 3, pp. 215-238). New York, NY: Garland Publishing.

Fenton, W. N. (1985). Another Eagle Dance for Gahehdagowa (FGS). In Tooker, E. (Ed.), An Iroquois Source Book: Medicine Society Rituals (Vol. 3, pp. 60-64). New York, NY: Garland Publishing.

Fenton, W. N. (1987). The False Faces of the Iroquois. Norman, OK: University of Oklahoma Press.

Fiddes, G. W. J. (1965). He Took Down His Shingle (A Backward Look at the Indian Medicine Man). Canadian Journal of Public Health, 56, 400-401.

Fields, S. (1976). Folk Healing for the Wounded Spirit. II. Medicine Men: Purveyors of an Ancient Art. Innovations, 3, 12-18.

Fienup Riordan, A. (1997). Present Yup'ik Recollections of Past Shamans. Etudes/Inuit/Studies, 21(1-2), 229-244.

Figueira, D. M. (1993). Gloria Flaherty, Shamanism and the Eighteenth Century. History of Religions, 33(2), 207-??

Fikes, J. C. (1992). Carlos Castaneda, Academic Opportunism, and the Psychedelic Sixties. Lanham, MD: Madison Press.

Filatov, S. (2000). Yakutia (Sakha) Faces a Religious Choice: Shamanism or Christianity. Religion, State & Society, 28(1), 113-122.

Firth, R. (1964). Shamanism. In J. Gould, and W. Kolb (Eds), Dictionary of the Social Sciences (pp. 638-639). New York: Free Press of Glencoe.

Fisken, R. A. (1997). Scientist or shaman? The Lancet, 9094, 1860.

Flaherty, G. (1992). Shamanism and the Eighteenth Century. Princeton, NJ: Princeton University Press.

Flaherty, G., & Fink, K. J. (1995). Shamanism and the Eighteenth Century. Comparative Literature Studies, 32(3), 2.

Fletcher, C. M., & Kirmayer, L. J. (1997). Spirit Work: Nunavimmiut Experiences of Affliction and Healing. Etudes/Inuit/Studies, 21(1-2), 189-208.

Florsheim, P. (1990). Cross-cultural Views of Self in the Treatment of Mental Illness: Disentangling the Curative Aspects of Myth from the Mythic Aspects of Cure. Psychiatry, 53(3), 304-15.

Fogelson, R. D. (1980). The Conjuror in Eastern Cherokee Society. Journal of Cherokee Studies, 5, 60-87.

Fox, J. R. (1964). Witchcraft and Clanship in Cochiti Therapy. In Kiev, A. (Ed.), Magic, Faith, and Healing (pp. 174-200). New York, NY: Free Press.

Fox, J. R. (1967). Witchcraft and Clanship in Cochiti Therapy. In Middleton, J. (Ed.), Magic, Witchcraft, and Curing (pp. 255-284). Garden City, NJ: Natural History Press.

Fox, M. J. T. (2002). Shamanism. The Social Science Journal, 39(Part 2), 314-316.

Francfort, H.-P., & Hamayon, R. (2002). The Concept of Shamanism, Uses and Abuses. La Recherche, 33(357), 4.

Franefort, H.-P., Hamayon, R. N., Bahn, P. G., & Charleux, I. (2003). Rezensionen - The Concept of Shamanism. Anthropos, 98(2), 3.

Frederiksen, S. (1966). The Structure and Function of the Soul in Eskimo Shamanism. Paper presented at the Congreso Internacional de Americanistas, 36th, Espana.

Freeman, D. (1966). Shaman and Incubus. In W. Muesterberger, and S. Axelrad (Eds), The Psychoanalytic Study of Society (pp. 315-343). New York: International Universities Press.

Frisbie, C. J. (1993). The Navajo Ceremonial Practitioners Registry. Journal of the Southwest, 35(1), 53-92.

Frison, G. C., & Van Norman, Z. (1993). Carved Steatite and Sandstone Tubes: Pipes for Smoking or Shaman's Paraphernalia. Plains Anthropologist, 38(143), 163-176.

Furst, P. T. (1974). The Roots and Continuities of Shamanism. Artscanada, 30, 33-60.

Furst, P. T. (1977). The Roots and Continuities of Shamanism. In A.T. Brodzky, R. D., and N. Johnson (Ed.), Stones, Bones, and Skin: Ritual and Shamanic Art (Vol. 1-28): Toronto: Society for Art Publications.

Furst, P. T. (1989). The Water of Life: Symbolism and Natural History on the Northwest Coast. Dialectical Anthropology, 14(2), 95-115.

Furst, P. T. (2004). Visionary Plants and Ecstatic Shamanism. Expedition, 46(1), 4.

Gaddis, V. H. (1977). American Indian Myths and Mysteries. Radnor, PA: Chilton Book Company.

Gale, A., & Gale, M. (19[n.d.]). Songs and Stories of the Red Man. Unpublished manuscript, Iowa City, IA.

Garrity, J. F. (2000). Jesus, Peyote, and the Holy People: Alcohol Abuse and the Ethos of Power in Navajo Healing. Medical Anthropology Quarterly, 14(4), 521-542.

Gaylord, S. (1999). Alternative Therapies and Empowerment of Older Women. Journal of Women & Aging, 11(2-3), 29.

Gayton, A. H. (1930). Yokuts-Mono Chiefs and Shamans. Berkeley, CA: University of California.

Gayton, A. H. (1976). Yokuts-Mono Chiefs and Shamans. In Bean, L. J. & Blackburn, T. C. (Eds.), Native Californians: A Theoretical Retrospective (pp. 175-223). Ramona, CA: Ballena Press.

Gearey, J. (1994). Gloria Flaherty, Shamanism in the Eighteenth Century. JEGP. Journal of English and Germanic Philology, 93(4), 614.

Gebhardt, K. H. (2003). Distinct Medical Treatment on the Positive List. This is Not Shamanism! MMW Fortschr Med, 145(22), 16.

Gebhardt, K. H. (2003). Science and Politics. Comment on the Article from DMW 15/2003. Dtsch Medical Wochenschr, 128(27), 1500.

Geertz, A. (1993). Archaic Ontology and White Shamanism. Religion, 23(4), 369-372.

Gernet, A. v. (2000). North American Indigenous 'Nicotiana' Use and Tobacco Shamanism: The Early Documentary Record, 1520-1660. In Winter, J. C. (Ed.), Tobacco Use by Native North Americans: Sacred Smoke and Silent Killer (pp. 59-80). Norman, OK: University of Oklahoma Press.

Gesner, A. (1884). Report of Warm Springs Agent. In Annual Report of the Commissioner of Indian Affairs to the Secretary of the Interior for the Year 1884 (pp. 150-153). Washington, DC: Government Printing Office.

Gesner, A. (1885). Report of Warm Springs Agent. In Annual Report of the Commissioner of Indian Affairs to the Secretary of the Interior for the Year 1885 (pp. 171-176). Washington, DC: Government Printing Office.

Ghoshal, S. C. (1993). Shamanism to Overcome Disaster: A Case Study. Man in India, 73(2), 137.

Gibson, E. M. (1872). No. 67: Annual Report of E.M. Gibson, Neah Bay Agency, Washington Territory. In Annual Report of the Commissioner of Indian Affairs to the Secretary of the Interior for the Year 1872 (pp. 350-352). Washington, DC: Government Printing Office.

Gill, S. D. (1974). Prayer of the Navajo Carved Figurine: An Interpretation of the Navajo Remaking Ritual. Plateau, 47, 59-69.

Gillham, C. E. (1955). Medicine Men of Hooper Bay. New York, NY: Macmillan.

Gillin, John. (1948). Magical Fright. Psychiatry, 11, 387-400.

Glaser, B. (2004). Ancient Traditions Within a New Drama Therapy Method: Shamanism and Developmental Transformations. The Arts in Psychotherapy, 31(2), 12.

Glass-Coffin, B. (1999). Engendering Peruvian Shamanism through Time: Insights from Ethnohistory and Ethnography. Ethnohistory, 46(2), 205-238.

Go, V. L., & Champaneria, M. C. (2002). The New World of Medicine: Prospecting for Health. Nippon Naika Gakkai Zasshi, 91, 159-163.

Goldman, C. (1994). Garden of Lost Souls. Yoga Journal(118), 80.

Goldwert, M. (1992). The Psychiatrist as Shaman: Sullivan and Schizophrenia. Psychological Reports, 70(2), 669-670.

Gonzalez, N. (1975). Pomo Shaman. American Anthropologist, 77, 177-178.

Goodman, L. J. (1989). Mescalero Apache Medicine Men: An Aid to Living a Fine Life. El Palacio, 95(1), 31-37.

Gottlieb, O. R., & Borin, M. R. (2002). Shamanism or science? Anais Academia Brasileira de Ciencias, 74(1), 135-144.

Grant, R. E. (1982). Tuuhikya: The Hopi Healer. American Indian Quarterly, 6, 283-290.

Grant, S. D. (1997). Religious Fanaticism at Leaf River, Ungava, 1931. Etudes/Inuit/Studies, 21(1-2), 159-188.

Gray, L. R. (1984). Healing Among Native American Indians. PSI Research, 3(3-4), 141-149.

Green, J. T. (1998). Near-Death Experiences, Shamanism, and the Scientific Method. Journal of Near-Death Studies, 16(3), 18.

Greer, M., & Greer, J. (2003). A test for Shamanic Trance in Central Montana Rock Art. Plains Anthropologist, 48(186), 105-120.

Greer, M., & Greer, J. (2004). Reply to Kehoe: Rock Art and Shamanism. Plains Anthropologist, 49(189), 81-84.

Grim, J. A. (1981). Reflections on Shamanism: The Tribal Healer and the Technological Trance. Chambersburg, PA: Anima Books.

Grim, J. A. (1983). The Shaman Patterns of Siberian and Ojibway Healing. Norman, OK: University of Oklahoma Press.

Grim, J. A. (1990). Schlesier, Karl H The Wolves of Heaven: Cheyenne Shamanism, Ceremonies, and Prehistoric Origins. American Indian Quarterly, XIV(3), 317-318.

Groesbeck, C. J. (1989). C. G. Jung and the Shaman's Vision. Journal of Analytical Psychology, 34(3), 255-75.

Grossinger, R. (1990). Planet Medicine: From Stone Age Shamanism to Post-Industrial Healing. Berkeley, CA: North Atlantic Books.

Gruber, E. (1982). Trance-Formation: Schamanismus und die Auflosung der Ordnung. Basel: Sphinx Verlag.

Guedon, M. F. (1974). Chamanisme Tsimshian et Athapaskan un essai sur la definition des methodes chamaniques. In Barkow, J. H. (Ed.), Proceedings of the First Congress, Canadian Ethnological Society (pp. 186-222). Ottawa, Ontario: National Museums of Canada.

Guedon, M. F. (1982). Problemes de definition du chamanisme chez les Amerindiens de la cote Nord-ouest: l'exemple de Tsimshian. Culture, 2(3), 129-143.

Guedon, M. F. (1983). Surfacing: Amerindian Themes and Shamanism. In Grace, S. E. & Weir, L. (Eds.), Margaret Atwood: Language, Text, and System (pp. 91-111). Vancouver, BC: University of British Columbia Press.

Guedon, M. F. (1984). An Introduction to Tsimshian Worldview and its Practitioners. In Seguin, M. (Ed.), The Tsimshian (pp. 137-159). Vancouver, BC: University of British Columbia Press.

Guedon, M. F. (1984). Tsimshian Shamanic Images. In Seguin, M. (Ed.), The Tsimshian (pp. 174-211). Vancouver, BC: University of British Columbia Press.

Guerrier, E. A. M. (2000). Ethnographic Inventions: The Construction and Commodification of the Shaman Through Anthropological Discourse. In Nichols, J. D. (Ed.), Papers of the Thirty-first Algonquian Conference (pp. 130-143). Winnipeg, Canada: University of Manitoba.

Gunn, S. W. A. (1966). Totemic Medicine and Shamanism Among the Northwest American Indians. Journal of the American Medical Association, 196, 700-706.

Guthrie, S. E., Castillo, R. J., Throop, C. J., Wright, P., & Douglas, M. (2004). Book Review Forum -- Michael Winkelman's book Shamanism The Neural Ecology of Consciousness and Healing - Westport, CT and London: Bergin and Garvey, 2000. Journal of Ritual Studies, 18(1), 96.

Haase, E. (1987). Der Schamanismus der Eskimos: 1st. [s.n.]: Aachen.

Haase, E. (1993). Masken und Schamanen bei den Eskimo. In Indianer Nordamerikas: Kunst und Mythos (pp. 102-121). Mainz: Verlag Hermann Schmidt.

Haeberlin, H. K. (1918). SbEtStdaq, A Shamanistic Performance of the Coast Salish. American Anthropologist, 20, 249-257.

Halifax, J. (1979). An Interview with Matsuwa. Many Smokes, 13(2), 10-11.

Halifax, J., ed. (1979). Shamanic Voices: A Survey of Visionary Narratives. New York, NY: E.P. Dutton.

Hamayon, R. N. (1994). The Eternal Return of the Everybody-for-Himself Shaman. Information Processing Letters, 52(1), 99.

Hand, W.D., ed. (1976). Shamanic Equilibrium: Balance and Mediation in Known and Unknown Worlds. American Folk Medicine: A Symposium. Berkeley: University of California Press.

Handelman, D. (1967). The Development of a Washo Shaman. Ethnology, VI(4), 444-464.

Handelman, D. (1967). Transcultural Shamanic Healing: A Washo Example. Ethnos, 32, 149-166.

Handelman, D. (1968). Shamanizing on an Empty Stomach. American Anthropologist, 70, 353-356.

Handelman, D. (1970). Shamans: Explanation and Innovation. American Anthropologist, 72, 1093-1094.

Handelman, D. (1972). Aspects of the Moral Compact of a Washo Shaman. Anthropological Quarterly, 45(2), 84-101.

Handelman, D. (1976). The Development of a Washo Shaman. In Bean, L. J. & Blackburn, T. C. (Eds.), Native Californians: A Theoretical Retrospective (pp. 379-405). Ramona, CA: Ballena Press.

Hanegraaff, W. J. (1998). New Age Religion and Western Culture: Esotericism in the Mirror of Secular Thought. New York, NY: State University of New York Press.

Hansen, K. G. (1997). Unveiling the Treasures Left by Svend Frederiksen. Etudes/Inuit/Studies, 21(1-2), 245-248.

Hardin, T. (2001). At Home in the World Travel Touches the Soul, It Challenges Preconceptions, You get Underneath the Surface Impressions, Whether You End Up Challenging the Energy of the Vortex, Visiting Ancient Mysteries, Watching a Shaman at Work, Getting Seriously Spooked, or, Best of All, Just Living the Life of Riley. Successful Meetings : SM, 50(Part 9), 85-94.

Harner, M. (1962). Jivaro Souls. American Anthropologist, 62(2), 258-272.

Harner, M. (1990). The Way of the Shaman. New York, NY: HarperCollins.

Harner, M. (2002). Special Focus: Transpersonal Perspectives on Terrorism - Notes from Shamanism. The Journal of Transpersonal Psychology, 34(1), 2.

Harner, M. (2002). Shamanic Healing: We Are Not Alone. Retrieved October 15, 2002, from http://www.shamanism.org/articles/857415539.htm

Harris, L. (1879). Miraculous Healing Among the Zunis. Juvenile Instructor, 14, 173-176.

Harris, W. R. (1973). Practice of Medicine and Surgery by the Canadian Tribes in Champlain's Time. In Carter, W. H. (Ed.), Medical Practices and Burial Customs of the North American Indians (pp. 29-55). London, Ontario: Namind Printers and Publishers.

Harwood, A. (1998). The Medicine Wheel and the Millennium. Nursing Times, 94(30), 62-4.

Hawk, R. (1980). "I am a 'Wicasa Wakan'". In Walker, J. R. (Ed.), Lakota Belief and Ritual (pp. 136-137). Lincoln, NE: University of Nebraska Press.

Heart, B., & Larkin, M. (1996). The Wind is My Mother: The Life and Teachings of a Native American Shaman. New York, NY: Clarkson Potter.

Heaton, J. M. (2000). On R. D. Laing: Style, Sorcery, Alienation. Psychoanalysis Review, 87(4), 511-26.

Hedberg, J. (1980). Den bla hjorten: ur en undersokning om huichol-indianernas satt att leva med naturen och om shamanernas magi. Stockholm: Forfatterforlage.

Hedges, K. (1983). The Shamanic Origins of Rock Art. In Tilburg, J. A. v. (Ed.), Ancient Images on Stone: Rock Art of the Californias (pp. 46-61). Los Angeles, CA: Rock Art Archive.

Heillig Morris, R. (1995). Woman as Shaman: Reclaiming the Power to Heal. Women's studies, 24(6), 573.

Heinrichs, H. J. (1977). Psychoanalysis and Shamanism. Research of Michel Leiris. Psyche (Stuttg), 31(5), 457-75.

Heinze, R.-I. (1982). Shamans or Mediums: Toward a Definition of Different States of Consciousness. Phoenix: Journal of Transpersonal Anthropology, 6, 25-44.

Henderson, E. B. (1982). Kaibeto Plateau Ceremonialists: 1860-1980. In Brugge, D. M. & Frisbie, C. J. (Eds.), Navajo Religion and Culture: Selected Views: Papers in Honor of Leland C. Wyman (pp. 164-175). Santa Fe, NM: Museum of New Mexico Press.

Hendricks, J. W. (1988). Power and Knowledge: Discourse and Ideological Transformation Among the Shuar. American Ethnologist, 15(2), 216-238.

Hes, J. P. (1975). Shamanism and Psychotherapy. Psychotherapy and Psychosomatic Medical Psychology, 25(1-6), 251-3.

Hewitt, J. N. B., & Swanton, J. R. (1987). Notes on the Creek Indians. In Sturtevant, W. C. (Ed.), A Creek Source Book (pp. 119-159). New York, NY: Garland Publishing.

Heye, G. G. (1927). Shaman's Cache from Southern California. Indian Notes, 4, 315-323.

Hieb, L. A. (2001). Hopi Stories of Witchcraft, Shamanism, and Magic By Ekkehart Malotki and Ken Gary. American Indian Culture and Research Journal, 25(Part 3), 195-196.

Hippler, A. E. (1971). Shaman Curers and Personality. Suggestions Toward a Theoretical Model. Transcultural Psychiatric Research, 8, 190-193.

Hobgood, J. (1963). The prayer of a Tepehuan Witch Doctor. Tlalocan, 4(3), 255-258.

Hobson, G. (2002). The Rise of the White Shaman: Twenty-Five Years Later. Studies in American Indian literatures: Newsletter of the Association for Study of American Indian Literatures, 14(Part 2/3), 1-11.

Hodgson, D. (2000). Shamanism, Phosphenes, and Early Art: An Alternative Synthesis. Current anthropology, 41(5), 7.

Hoff, J. R. (1999). Colliding of Cultures in a Therapeutic Regimen. Journal of Cultural Diversity, 6(1), 3-4.

Hoffman, W. J. (1888). Pictography and Shamanistic Rites of the Ojibwa. American Anthropologist, 1, 209-229.

Honigmann, J. J. (1949). Parallels in the Development of Shamanism Among Northern and Southern Athapaskans. American Anthropologist, 51, 512-514.

Hoover, R. L. (1975). Chumash Sunsticks. Masterkey, 49, 105-109.

Hoppal, M. (1989). Changing Image of the Eurasian Shamans. In Hoppal, M. & Von Sadovszky, O. J. (Eds.), Shamanism: Past and Present (Vol. 1, pp. 75-90). Budapest, Hungary: International Society for Trans-Oceanic Research.

Hoppal, M. (1998). Shamanism: Selected Writings of Vilmos Dioszegi. Budapest: Akademiai Kiado.

Hoskins, J. (1988). The Drum is the Shaman, the Spear Guides His Voice. Social Science Medicine, 27(8), 819-28.

Howard, J. H. (1974). The Arikara Buffalo Society Medicine Bundle. Plains Anthropologist, 19(66), 241-271.

Hrdlicka, A. (1943). Alaska Diary, 1926-1931. Lancaster, PA: Jacques Cattell Press.

Hudson, D. T. (1979). A Rare Account of Gabrielino Shamanism from the Notes of John P. Harrington. Journal of California and Great Basin Anthropology, 1, 356-362.

Hultgren, G. M., & Jeffers, J. S. (1994). Shamanism, a Religious Paradigm: Its Intrusion into the Practice of Chiropractic. Journal of Manipulative and Physiological Therapeutics, 17(6), 404.

Hultkrantz, A. (1985). The Shaman and the Medicine-Man. Social Science & Medicine, 20(5), 511-5.

Hultkrantz, A. (1989). The Place of Shamanism in the History of Religions. In Hoppal, M. & Von Sadovszky, O. J. (Eds.), Shamanism: Past and Present (pp. 43-52). Budapest, Hungary: International Society for Trans-Oceanic Research.

Hultkrantz, A. (1992). Shamanic Healing and Ritual Drama: Health and Medicine in Native North American Religious Traditions. New York, NY: Crossroad.

Hultkrantz, A. (1997). Shamanic Healing and Ritual Drama. New York, NY: Crossroad Publishing Company.

Icamina, P. (1993). Threads of Common Knowledge. IDRC Rep, 21(1), 14-6.

Ingerman, S. (2003). Sandra Ingerman, MA. Medicine for the Earth, Medicine for People. Interview by Bonnie Horrigan. Alternative Therapeutic Health Medicine, 9(6), 76-84.

Ipellie, A. (1974). The Midnight Shaman. Inuit Monthly, 3(9), 86.

Irimoto, T. (1990). Interpreting Shamanism and Folk Religion in Arctic and Subarctic Regions. Minzokugaku-Kenkyu Japanese journal of Ethnology, 55(2), 230-231.

Irwin, L. (1992). Cherokee Healing: Myth, Dreams, and Medicine. American Indian Quarterly, 16(2), 237-257.

Irwin, L. (1994). The Dream Seekers: Native American Visionary Traditions of the Great Plains. Norman, OK: University of Oklahoma Press.

Istomin Durante, I. (1974). L'uomo Medicina fra i popoli del Nord-America. Terra Ameriga, 10(31-32), 59-68.

Jackson, J. (1995). Preserving Indian Culture: Shaman Schools and Ethno-Education in the Vaupes, Columbia. Cultural Anthropology, 10(3), 302-329.

Jackson, S. W. (2001). The Wounded Healer. Bulletin of Historical Medicine, 75(1), 1-36.

Jakobsen, M. D. (1999). Shamanism: Traditional and Contemporary Approaches to the Mastery of Spirits and Healing. New York, NY: Berghahn Books.

James, J. S. (1999). Diarrhea: New Treatment Option from Shaman. AIDS Treat News(No 324), 1-2.

Jarvenpa, R. (1985). Northern Pilgrimage. Beaver, 315(4), 54-59.

Jett, S. C. (1991). Pete Price, Navajo Medicineman (1868-1951): A Brief Biography. American Indian Quarterly, 15, 91-103.

Jette, J. (1907). On the Medicine-Men of the Ten'a. Jounral of the Royal Anthropological Institute of Great Britain and Ireland, 37, 157-188.

Jilek, W. G. (1974). Indian Healing Power: Indigenous Therapeutic Practices in the Pacific Northwest. Psychiatric Annals, 4(11), 13-21.

Jilek, W. G. (1982). Indian Healing: Shamanic Ceremonialism in the Pacific Northwest Today: Blaine, WA: Hancock House Publishing.

Jilek, W. G. (1982). Indian Healing: Shamanic Ceremonialism in the Pacific Northwest Today. Surrey, BC: Hancock House Publishers.

Jilek, W. G., & Jilek Aall, L. (1978). The Psychiatrist and His Shaman Colleague: Cross-Cultural Collaboration with Traditional Amerindian Therapists. Journal of Operational Psychiatry, 9, 32-39.

Jilek, W. G., & Jilek Aall, L. (1982). Shamanic Symbolism in Salish Indian Rituals. In The Logic of Culture (pp. 127-136). South Hadley, MA: J.F. Bergin Publishers.

Jilek, W. G., Jilek Aall, L., Todd, N., & Galloway, B. (1979). Symbolic Processes in Contemporary Salish Indian Ceremonials. In Ridington, R. (Ed.), Special Issue on Symbolic Anthropology (pp. 36-57). Edmonton, Alberta: Department of Anthropology, University of Alberta.

Jilek, W. G., & Todd, N. (1974). Witchdoctors Succeed Where Doctors Fail: Psychotherapy Among Coast Salish Indians. Canadian Psychiatric Association Journal, 19, 3551-3556.

Johansen, J. P. (1940). Den Ostgronlandske angakoqkult og dens Sorudsaetninger. Geografisk Tidsskrift, 43, 31-55.

Johnson, B. (2000). Plastic Shaman in the Global Village: Understanding Media in Thomas King's Green Grass, Running Water. Studies in Canadian Literature, 25(Part 2), 24-49.

Johnson, C. L. (1981). Psychoanalysis, Shamanism and Cultural Phenomena. Journal of the Academy of Psychoanalysis, 9(2), 311-318.

Johnson, F. (1943). Notes on Micmac Shamanism. Primitive Man, 16, 53-80.

Johnson, P. (1995). Shamanism from Ecuador to Chicago: A Case Study in New Age Ritual Appropriation. Religion, 25(2), 16.

Johnson, R. (1973). The Art of the Shaman. Iowa City, IA: University of Iowa Museum of Art.

Johnson, W. (1993). The Visions of Luciano Perez, Contemporary Native American Shaman. Religion, 23(4), 343-354.

Johnson, W. (1994). An Unusual Transmission of the Lakota Yuwipi Ritual. Journal of Ritual Studies, 8(1), 95-124.

Jolly, A. (1999). Indigenous Mental Health Care Among Gurkha Soldiers Based in the United Kingdom. J R Army Med Corps, 145(1), 15-7.

Jonaitis, A. C. (1978). Land Otters and Shamans: Some Interpretations of Tlingit Charms. American Indian Art, 4(1), 62-66.

Jonaitis, A. C. (1981). Creations of Mystics and Philosophers: The White Man's Perceptions of Northwest Coast Indian Art from the 1930s to the Present. American Indian Culture and Research Journal, 5(1), 1-45.

Jonaitis, A. C. (1982). Sacred Art and Spiritual Power: An Analysis of Tlingit Shaman's Masks. In Mathews Pearlstone, Z. & Jonaitis, A. (Eds.), Native North American Art History: Selected Readings (pp. 119-136). Palo Alto, CA: Peek Publications.

Jonaitis, A. C. (1983). Liminality and Incorporation in the Art of the Tlingit Shaman. American Indian Quarterly, 7(3), 41-68.

Jonaitis, A. C. (1983). Style and Meaning in the Shamanic Art of the Northern Northwest Coast. In Holm, B. (Ed.), The Box of Daylight: Northwest Coast Indian Art (pp. 129-131). Seattle, WA: Seattle Art Museum and University of Washington Press.

Jones, D. E. (1984). Sanapia: Comanche Medicine Woman. Prospect Heights, IL: Waveland Press.

Jones, F. (1997). Last of Wintu Doctors. In Crozier-Hogle, L. & Wilson, D. B. (Eds.), Surviving in Two Worlds: Contemporary Native American Voices (pp. 21-29). Austin, TX: University of Texas Press.

Jones, Peter N. (2004). Shamanism in North America: A Comprehensive Bibliography on the Use of the Term. Boulder, CO: Bauu Institute Press.

Jones, Peter N. (2006). Shamanism: An Inquiry into the History of the Scholarly Use of the Term in English-Speaking North America. Anthropology of Consciousness, 17(2): 4-32.

Joralemon, D. (1990). The Selling of the Shaman and the Problem of Informant Legitimacy. Journal of Anthropological Research, 46(2), 105.

Judd, N. M. (1947). When the Jemez Medicine Men came to Zuni. Journal of American Folklore, 40, 182-184.

Kan, S. (1991). Shamanism and Christianity: Modern-Day Tlingit Elders Look at the Past. Ethnohistory, 38(4), 363-387.

Karafistan, R. (2003). "The Spirits Wouldn't Let Me Be Anything Else": Shamanic Dimensions in Theatre Practice Today - "Third Theatre" and its Relationship with the Role of the Shaman. New Theatre Quarterly : NTQ, 19(74), 19.

Kasten, E. (1989). Sami Shamanism from a Diachronic Point of View. In Hoppal, M. & Von Sadovszky, O. J. (Eds.), Shamanism: Past and Present (Vol. 1, pp. 115-123). Budapest, Hungary: International Society for Trans-Oceanic Research.

Kehoe, A. B. (2000). Shamans and Religion: An Anthropological Exploration in Critical Thinking. Prospect Heights, IL: Waveland Press.

Kehoe, A. B. (2004). Testing for "Shamanic Trance" in Rock Art: A Comment on Greer and Greer. Plains Anthropologist, 49(189), 79-80.

Kelly, I. T. (1936). Chemehuevi Shamanism. In Essays in Anthropology Presented to Alfred Louis Kroeber (pp. 129-142). Berkeley, CA.

Kelly, I. T. (1939). Southern Paiute Shamanism. Berkeley, CA: University of California Press.

Kemnitzer, L. S. (1976). Structure, Content, and Cultural Meaning of "Yuwipi": A Modern Lakota Healing Ritual. American Ethnologist, 3, 261-280.

Kemnitzer, L. S. (1977). How a Shamanic Ritual Helps Organize a Difuse Political Revitalization Movement. Ethnomedicine, Ethnobotany and Ethnozoology, 4(1-2), 115-126.

Kemnitzer, L. S. (1978). Yuwipi. Indian Historian, 11(2), 2-5.

Keppler, J. (1985). Comments on Certain Iroquois Masks. In Tooker, E. (Ed.), An Iroquois Source Book: Medicine Society Rituals (Vol. 3, pp. 1-40). New York, NY: Garland Publishing.

Kew, M., & Kew, D. (1981). "People Need Friends, It Makes Their Minds Strong": A Coast Salish Curing Rite. In Abbott, D. N. (Ed.), The World is as Sharp as a Knife: An Anthology in Honour of Wilson Duff (pp. 29-36). Victoria, BC: British Columbia Provincial Museum.

King, A. G. (1960). Shamanism. Obstetrics and Gynecology, 16, 129-132.

King, S. R. (2003). Biocultural Diversity, Phytomedicines, and Tropical Rainforests: The Holistic Link from Practitioner to Cultures of the Tropical Rainforest. Journal of Alternative and Complementry Medicine, 9(6), 813-815.

King, S. R., & Carlson, T. J. (1995). Biocultural Diversity, Biomedicine and Ethnobotany: The Experience of Shaman Pharmaceuticals. Interciencia, 20(3), 6.

King, S. R., Carlson, T. J., & Moran, K. (1996). Biological Diversity, Indigenous Knowledge, Drug Discovery and Intellectual Property Rights: Creating Reciprocity and Maintaining Relationships. Journal of Ethnopharmacology, 51(1-3), 45-57.

King, S. R., & Tempesta, M. S. (1994). From Shaman to Human Clinical Trials: The Role of Industry in Ethnobotany, Conservation and Community Reciprocity. Ciba Foundation Symposium(185), 197.

Kirmayer, L. J. (2003). Asklepian Dreams: The Ethos of the Wounded-Healer in the Clinical Encounter. Transcultural Psychiatry, 40(2), 248-277.

Kleivan, I., & Sonne, B. (1985). Eskimos: Greenland and Canada. Leiden: E.J. Brill.

Klopfer, B., & Boyer, L. B. (1961). Notes on the Personality Structure of a North American Indian Shaman Rorschach Interpretation. Journal of Projective Techniques and Personality Assessment, 25, 170-178.

Kniffen, F. B., Gregory, H. F., & Stokes, G. A. (1987). The Historic Indians Tribes of Louisiana: From 1542 to the Present. Baton Rouge, LA: Louisiana State University Press.

Knudtson, P. M. (1975). Flora, Shaman of the Wintu. Natural History, 84, 12-13.

Knudtson, P. M. (1991). Flora, Shaman of the Wintu. In Bean, L. J. & Vane, S. B. (Eds.), Ethnology of the Alta California Indians: Postcontact (Vol. 2, pp. 355-359). New York, NY: Garland Publishing.

Koenigsberger, D. (1993). Shamanism and the Eighteenth Century, Gloria Flaherty. History of European Ideas, 17(2-3), 354-354.

Koepping, E. P. (1976). On the Epistemology of Participant Observation and the Generating of Paradigms: Some Critical Reflections on Victor Turner, Carlos Castaneda and Applied Shamanism. Occasional Papers of the University of Queensland Anthropological Museum, 6, 159-177.

Kokan, S. (1995). Shamanism: Past and Present Edited by Mihaly Hoppal and Otto J. von Sadovszky. Asian Folklore Studies, 54(1), 135.

Kostash, M. (1997). The Shaman. Chatelaine, 70(4), 113.

Kracke, W. H. (1987). "Everyone Who Dreams Has a Bit of Shaman": Cultural and Personal Meanings of Dreams - Evidence From the Amazon. Psychiatric Journal of the University of Ottawa, 12(2), 65-72.

Kraus, R. F. (1972). A Psychoanalytic Interpretation of Shamanism. Psychoanalytic Review, 59(1), 19-32.

Kraut, A. M. (1990). Healers and Strangers. Immigrant Attitudes Toward the Physician in America - A Relationship in Historical Perspective. Jama, 263(13), 1807-1811.

Krech, S. I. (1981). "Throwing Bad Medicine": Sorcery, Disease, and the Fur Trade Among the Kutchin and other Northern Athapaskans. In Krech, S. I. (Ed.), Indians, Animals, and the Fur Trade: A Critique of "Keepers of the Game" (pp. 75-108). Athens, GA: University of Georgia Press.

Krippner, S. (1987). Shamanism, Personal Mythology, and Behavior Change. International Journal of Psychosomatics, 34(4), 22-7.

Krippner, S. (1992). The Shaman as Healer and Psychotherapist. Voices; the Art and Science of Psychotherapy, 28(4), 12.

Krippner, S. (1997). Stanley Krippner, PhD. Medicine and the Inner Realities. Interview by Bonnie Horrigan. Alternative Therapies and Health Medicine, 3(6), 98-107.

Krippner, S. (2000). The Epistemology and Technologies of Shamanic States of Consciousness, Journal of Consciousness Studies, 7(11-12), 93-118.

Krippner, S., and Allen Combs. (2002). The Neurophenomenology of Shamanism, Journal of Consciousness Studies, 9(3), 77-82.

Krippner, S. (2002). Conflicting Perspectives on Shamans and Shamanism: Points and Counterpoints. American Psychologist, 57(11), 960-977.

Krippner, S., & Combs, A. (2002). The Neurophenomenology of Shamanism: An Essay Review. Journal of Consciousness Studies, 9(3), 77-82.

Krippner, S., & Combs, A. (2002). The Neurophenomenology of Shamanism. Re-vision, 24(Part 3), 46-48.

Krippner, S., & Thompson, A. (1996). A 10-Facet Model of Dreaming Applied to Dream Practices of Sixteen Native American Cultural Groups. Dreaming, 6(2), 71-96.

Kugel, R. (1994). Of Missionaries and Their Cattle: Ojibwa Perceptions of a Missionary as Evil Shaman. Ethnohistory, 41(2), 227-244.

Kugel, R. (2000). Of Missionaries and Their Cattle: Ojibwa Perceptions of a Missionary as Evil Shaman. In Mancall, P. C. & Merrell, J. H. (Eds.), American Encounters: Natives and Newcomers from European Contact to Indian Removal, 1500-1850 (pp. 161-175). New York, NY: Routledge.

Kurath, G. P. (1959). Blackrobe and Shaman. [s.n.]: Michigan Academy of Science, Arts and Letters.

La Barre, W. (1947). Kiowa Folk Sciences. The Journal of American Folklore, 60(236), 105-114.

La Barre, W. (1972). Hallucinogens and the Shamanic Origins of Religion. In P.T. Furst, Ed., Flesh of the Gods (pp. 261-278). New York, NY: Praeger.

La Farge, P. (1976). Huichol Shamanic Religion and its Artifacts. El Palacio, 82(1), 9-16.

La Flesche, F. (1890). The Omaha Buffalo Medicine-Men. Journal of American Folklore, 3, 215-221.

La Flesche, F. (1904). Who Was the Medicine Man? Unpublished Manuscript, Omaha, NE.

Lafleur, L. J. (1940). On the Mide of the Ojibway. American Anthropologist, 42, 705-707.

Laird, C. (1976). The Chemehuevis. Banning, CA: Malki Museum Press.

Laird, C. (1980). Chemehuevi Shamanism, Sorcery, and Charms. Journal of California and Great Basin Anthropolgy, 2(1), 80-7.

Lake, R. G., Jr. (1983). Shamanism in Northwestern California: A Female Perspective on Sickness, Healing and Health. White Cloud Journal, 3(1), 31-42.

Lake, R. G., Jr. (1991). Native Healer: Initiation into an Ancient Art. Wheaton, IL: Quest Books.

Lake, T. S. H. (1996). Hawk Woman Dancing with the Moon. New York, NY: M. Evans.

Lamb, F. B. (1985). Rio Tigre and Beyond: The Amazon Jungle Medicine of Manuel Cordova. Berkeley, CA: North Atlantic Books.

Lame Deer, A. F. (1992). Gift of Power: The Life and Teachings of a Lakota Medicine Man. Santa Fe, NM: Bear & Co.

Lamphere, L. (2000). Comments on the Navajo Healing Project. Medical Anthropology Quarterly, 14(4), 598-602.

Langdon, E. J. (1989). Shamanism as the History of Anthropology. In Hoppal, M. & Von Sadovszky, O. J. (Eds.), Shamanism: Past and Present (Vol. 1, pp. 53-68). Budapest, Hungary: International Society for Trans-Oceanic Research.

Langenwalter, R. E. (1980). A Possible Shaman's Cache from CA-Riv-102, Hemet, California. Journal of California and Great Basin Anthropology, 2, 233-244.

Large, R. G. (1968). Drums and Scalpel: From Native Healers to Physicians on the North Pacific Coast. Vancouver, BC: Mitchell Press.

Laugrand, F. (1997). 'Le siqqitiq': renouvellement religieux et premier rituel de conversion chez les Inuit de nnord de la Terre de Baffin. Etudes/Inuit/Studies, 21(1-2), 101-140.

Layard, J. W. (1930). Shamanism: An Analysis Based on Comparison with the Flying Tricksters of Malekula. Journal of the Royal Anthropological Institute of Great Britain and Ireland, 60, 525-550.

Layton, R. (2000). Shamanism, Totemism and Rock Art: Les Chamanes de la Prâehistoire in the Context of Rock Art Research. Cambridge Archaeological Journal, 10(1), 18.

Lee, G. (1980). An Unusual Carved Figure from the Chumash Area. Journal of California and Great Basin Anthropology, 2, 263-266.

Lee, R., & Balick, M. J. (2002). Snakebite, Shamanism, and Modern Medicine: Exploring the Power of the Mind-Body Relationship in Healing. Alternative Therapies, 8(3), 118-121.

Lee, R., & Balick, M. J. (2003). Stealing the Soul, Soumwahu en Naniak, and Susto: Understanding Culturally-Specific Illnesses, Their Origins and Treatment. Alternative Therapies in Health Medicine, 9(3), 106-109.

Leh, L. L. (1934). The Shaman in Aboriginal American Society. In Anthropological Studies (Vol. 20, pp. 199-263). Boulder, Co: University of Colorado.

Lehmann, A. C., & Myers, J. E. (Eds.). (1997). Magic, Witchcraft, and Religion: An Anthropological Study of the Supernatural. Mountain View, CA: Mayfield Publishing Company.

Leighton, A. H., & Leighton, D. C. (1949). Gregorio, the Hand-Trembler. Cambridge, MA: Harvard University, Peabody Museum of American Archaeology and Ethnology.

Leighton, A.H., & Leighton, D.C. (1941). Elements of Psychotherapy in Navaho Religion. Psychiatry, 4, 515-523.

Levi, L. M. (1973). Il significato del Trickster in American settentrionale. Paper presented at the Congresso Internazionale Degli Americanisti, Rome, Casa Editrice Tilgher.

Levine, G. S. (1980). Rites and Customs: Healing Practices-Ritual. In Levine, G. S. (Ed.), Languages and Lore of the Long Island Indians (pp. 285-295). Stony Brook, NY: Suffolk County Archaeological Association.

Levi-Strauss, C. (1967). The Sorcerer and His Magic. In Middleton, J. (Ed.), Magic, Witchcraft, and Curing (pp. 23-42). New York, NY: American Museum of Natural History.

Levy, J. E. (1994). Hopi Shamanism: A Reappraisal. In DeMallie, R. J. & Ortiz, A. (Eds.), North American Indian Anthropology: Essays on Society and Culture (pp. 307-327). Norman, OK: University of Oklahoma Press.

Lewis, I. M. (1989). Ecstatic Religion: A Study of Shamanism and Spirit Possession. New York, NY: Routledge.

Lewis, I. M. (1994). Flaherty, G Shamanism and the Eighteenth Century. Man, 29(4), 996.

Lewis, J. R. (1988). Shamans and Prophets: Continuities and Discontinuities in Native American New Religions. American Indian Quarterly, 12(3), 221-228.

Lewis, T. H. (1974). An Indian Healer's Preventive Medicine Procedure. Hospital and Community Psychiatry, 25(2), 94-95.

Lewis, T. H. (1977). Therapeutic Techniques of Huichole Curanderos with a Case Report of Cross Cultural Psychotherapy (Mexico). Anthropos, 72, 709-716.

Lewis, T. H. (1980). The Changing Practice of the Oglala Medicine Man. Plains Anthropologist, 25(89), 265-267.

Lewis, T. H. (1987). The Contemporary Yuwipi. In DeMallie, R. J. (Ed.), Sioux Indian Religion (pp. 173-187). Norman, OK: University of Oklahoma Press.

Lewis, T. H. (1990). The Medicine Men: Oglala Sioux Ceremony and Healing. Lincoln, NE: University of Nebraska Press.

Lewis-Williams, J. D. (2003). Debate - Putting the Record Straight: Rock Art and Shamanism. Antiquity, 77(295), 5.

Lewis-Williams, J. D. (2003). Putting the Record Straight: Rock Art and Shamanism. Antiquity, 77(295), 5.

Lewis-Williams, J. D., Klein, C. F., & Stanfield-Mazzi, M. (2004). On Sharpness and Scholarship in the Debate on "Shamanism". Current Anthropology, 45(3), 2.

Lewton, E. L., & Bydone, V. (2000). Identity and Healing in Three Navajo Religious Traditions: Sa'ah Naaghai Bik'eh Hozh. Medical Anthropological Quarterly, 14(4), 476-97.

Lex, B. (1984). Recent Contributions to the Study of Ritual Trance. Reviews in Anthropology, 11, 44-51.

Lex, B. W., & Isaacs, H. L. (1980). Handling Fire: Treatment of Illness by the Iroquois False-Face Medicine Society. In Bonvillain, N. (Ed.), Studies on Iroquoian Culture (pp. 5-13). Rindge, NH: Department of Anthropology, Franklin Pierce College.

Liberty, M. P. (1970). Priest and Shaman on the Plains: A False Dichotomy? Plains Anthropologist, 15, 73-79.

Lindemann, M. (1993). Flaherty, Shamanism and the Eighteenth Century. Eighteenth-century Studies, 27(1), 111-113.

Linderman, F. B. (1972). Pretty-Shield, Medicine Woman of the Crows. New York, NY: John Day.

Linderman, F. B. (1974). Pretty-Shield: Medicine Woman of the Crows. Lincoln, NE: University of Nebraska Press.

Linton, -. R. (1923). Annual Ceremony of the Pawnee Medicine Men. Chicago, IL: Field Museum of Natural History.

Littlewood, R. (1989). Science, Shamanism and Hermeneutics: Recent Writing on Psychoanalysis. Anthropology Today, 5(1), 5-11.

Locke, R. G., & Kelly, E. F. (1985). A Preliminary Model for the Cross-Cultural Analysis of Altered States of Consciousness. Ethos, 13(1), 3-55.

Loeb, E.M. (1924). The Shaman of Nieu. American Anthropologist, 26, 393-402.

Loeb, E.M. (1929). Shaman and Seer. Amerian Anthropologist, 31, 60-84.

Lowie, R. H. (1925). A Trial of Shamans. In Parsons, E. C. (Ed.), American Indian Life (pp. 41-43). New York, NY: [s.n.].

Lucas, R. H., & Barrett, R. J. (1995). Interpreting Culture and Psychopathology: Primitivist Themes in Cross-Cultural Debate. Culture, Medicine, and Psychiatry, 19(3), 287-326.

Luckert, K. W., & Cook, J. C. (1979). Coyoteway: A Navajo Holyway Healing Ceremonial. Tucson, AZ: University of Arizona Press.

Luckey, J. C. (1892). Report of Warm Springs Agency. In Sixty-First Annual Report of the Commissioner of Indian Affairs to the Secretary of the Interior (pp. 422-426). Washington, DC: Government Printing Office.

Luer, G. M. (1993). A Safety Harbor Incised Bottle with Effigy Bird Feet and Human Hands from a Possible Headman Burial, Sarasota County, Florida. The Florida Anthropologist, 46(4), 238-248.

Luhrmann, T. M., & Jakobsen, M. D. (2001). Shamanism - Traditional and Contemporary Approaches to the Mastery of Spirits and Healing. TLS, the Times Literary Supplement(5108), 3.

Lynch, R. N. (1985). Seeing Twice: Shamanism, Berdache, and Homoeroticism in American Indian Culture. Southern Exposure, 13(6), 90-93.

Lyon, W. S. (Ed.). (1998). Encyclopedia of Native American Shamanism: Sacred Ceremonies of North America. Santa Barbara, CA: ABC-Clio.

MacDonald, G. F., Cove, J. L., Laughlin, C. D. J., & McManus, J. (1989). Mirrors, Portals, and Multiple Realities. Zygon, 24(1), 39-64.

Maclean, J. (1961). Blackfoot Medical Priesthood. Alberta Historical Review, 9(2), 1-7.

Manie, E. (2003). The shaman Reborn in Cyberspace, or Evolving Magico-Spiritual Techniques of Consciousness-Making. Technoetic Arts: An International Journal of Speculative Research, 1(1), 1-82.

Manyam, B. V., & Sanchez-Ramos, J. R. (1999). Traditional and Complementary Therapies in Parkinson's Disease. Advances in Neurology, 80, 565-574.

Maquet, J. (1978). Castaneda: Warrior or Scholar? American Anthropologist, 80, 362-363.

Margetts, E. L. (1975). Canadian Indian and Eskimo Medicine, with Notes on the Early History of Psychiatry Among the French and British Colonists. In Howells, J. G. (Ed.), World History of Psychiatry (pp. 400-431). New York, NY: Bruner/Mazel.

Mathews, Z. P. (1976). Huron Pipes and Iroquoian Shamanism. Man in the Northeast, 12, 15-31.

Matthews, W. (1888). The Prayer of a Navajo Shaman. American Anthropologist, 1, 149-170.

Maurer, R. L., Sr., Kumar, V. K., Woodside, L., & Pekala, R. J. (1997). Phenomenological Experience in Response to Monotonous Drumming and Hypnotizability. American Journal of Clinical Hypnosis, 40(2), 130-145.

Maybury-Lewis, D. (1991). Michael Taussig: Shamanism, Colonialism, and the Wild Man: A Study in Terror and Healing. Contemporary Sociology, 20(3), 375-376.

Mayo, L. (1991). Appropriation and the Plastic Shaman. Canadian Theatre Review, 54-55.

McAllister, J. G. (1970). Daveko: Kiowa-Apache Medicine Man. Austin, TX: Texas Memorial Museum.

McClellan, C. (1956). Shamanistic Syncretism in Southern Yukon. New York, NY: New York Academy of Sciences.

McClenon, J. (1993). The Experiential Foundations of Shamanic Healing. Journal of Medical Philosophies, 18(2), 107-127.

McClenon, J. (1997). Shamanic Healing, Human Evolution, and the Origin of Religion. Journal for the Scientific Study of Religion, 36(3), 345-354.

McGee, A. D. (1978). No Shame for the shaman. Canadian Nurse, 74(11), 22-23.

McIntosh, I. S. (2003). Seeking the Shaman. Cultural Survival Quarterly, 27(Part 2), 5-6.

McMurry, R. N. (19[n.d.]). Popular Lectures on Applied Psychology, Streamlining Your Personality, Maintaining Mental Health, Vocational Guidance, Saftey. Unpublished Manuscript, Iowa City, IA.

Mead, R. (1996, May 06). Selling Donna. New York Magazine, 29, 30.

Medicine Crow, J. (1979). From M.M. to M.D.: Medicine Man to Doctor of Medicine. Unpublished Manuscript, Bozeman, MT.

Medicine Crow, J. (1995). Indian Healing Arts. In Hirschfelder, A. (Ed.), Native Heritage: Personal Accounts by American Indians, 1790 to the Present (pp. 229-231). New York, NY: Macmillan.

Mehta, B. (1996). Shaman Woman. Journal of Caribbean Studies, 11(3), 193.

Meigs, P., III. (1972). Notes on La Huerta Jat'am, Baja California: Place Names, Hunting, and Shamans. Pacific Coast Archaeological Society Quarterly, 8(1), 35-40.

Merkur, D. (1983). Breath-Soul and Wind Owner: The Many and the One in Inuit Religion. American Indian Quarterly, 7(3), 23-39.

Merkur, D. (1985). Becoming Half Hidden: Shamanism and Initiation Among the Inuit. Stockholm: Almqvist & Wiksell International.

Merkur, D. (1992). Becoming Half Hidden: Shamanism and Initiation Among the Inuit. New York, NY: Garland.

Metzner, R. (1998). Hallucinogenic Drugs and Plants in Psychotherapy and Shamanism. Journal of Psychoactive Drugs, 30(4), 10.

Mikhailovskii, V. M. (1895). Shamanism in Siberia and European Russia, Being the Second Part of "Shamanstvo". Journal of the Anthropological Institute of Great Britain and Ireland, 24, 62-100.

Mikhailovskii, V. M. (1895). Shamanism in Siberia and European Russia- (Continued). Journal of the Anthropological Institute of Great Britain and Ireland, 24, 124-158.

Miller, C. L. (1996). Smohalla: (c. 1850-95) Wanapam Shaman and Prophet. In Hoxie, F. E. (Ed.), Encyclopedia of North American Indians (pp. 600-603). Boston, MA: Houghton Mifflin Company.

Miller, J. (1984). Tsimshian Religion in Historical Perspective: Shamans, Prophets, and Christ. In Miller, J. & Eastman, C. M. (Eds.), The Tsimshian and Their Neighbors of the North Pacific Coast (pp. 137-147). Seattle, WA: University of Washington Press.

Miller, J. (1985). Shamans and Power in Western North America: The Numic, Salish, and Keres. In Blackburn, T. C. (Ed.), Woman, Poet, Scientist: Essays in New World Anthropology Honoring Dr. Emma Louise Davis (pp. 56-66): Ballena Press: California.

Miller, J. (1996). Changing Moons: A History of Caddo Religion. Plains Anthropologist, 41(157), 243-259.

Miller, J. (1999). Lushootseed Culture and the Shamanic Odyssey: An Anchored Radiance. Lincoln, NE: University of Nebraska Press.

Mills, A., & Champion, L. (1996). Come-Backs/Reincarnation as Integration; Adoption-Out as Disassociation: Examples from First Nations Northwest British Columbia. Anthropology of Consciousness, 7(3), 30-43.

Mills, A., & Clifton-Percival, B. (2003). Shamanism Defends a People Halaits in the Gitxsan Fight for Rights. Cultural Survival Quarterly, 27(Part 2), 29-31.

Milne, D., & Howard, W. (2000). Rethinking the Role of Diagnosis in Navajo Religious Healing. Medical Anthropology Quarterly, 14(4), 543-70.

Mohatt, G. V., & Eagle Elk, J. (2000). The Price of a Gift: A Lakota Healer's Story. Lincoln, NE: University of Nebraska Press.

Moises, R., Kelley, J. H., & Holden, W. C. (1971). A Yaqui Life: The Personal Chronicle of a Yaqui Indian. Lincoln, NE: University of Nebraska Press.

Money, M. (1997). Shamanism and Complementary Therapy. Complementary Therapies in Nursing and Midwifery, 3(5), 131-135.

Money, M. (2000). Shamanism and Complementary Therapy. Complementary Therapies in Nursing and Midwifery, 6(4), 207-212.

Money, M. (2001). Shamanism as a Healing Paradigm for Complementary Therapy. Complementary Therapies in Nursing and Midwifery, 7(3), 6.

Mooney, J. (1982). Cherokee Theory and Practice of Medicine. Journal of Cherokee Studies, 7, 25-29.

Mooney, J. (1986). The Sacred Formulas of the Cherokees. In Ford, R. I. (Ed.), An Ethnobiology Sourcebook (pp. 301-397). New York, NY: Garland Publishing.

Morejohn, G. V., & Galloway, J. P. (1983). Identification of Avian and Mammalian Species Used in the Manufacture of Bone Whistles Recovered from a San Francisco Bay Area Archaeological Site. Journal of California and Great Basin Anthropology, 5, 87-97.

Morgan, W. (1970). Human-Wolves Among the Navaho. New Haven, CT: Human Relations Area Files Press.

Mousalimas, S. A. (1995). The Transition from Shamanism to Russian Orthodoxy in Alaska. Providence, RI: Berghahn Books.

Muller, K. E. (1998). Shamans as Healers. Krankenpfl Journal, 36(6), 245-246.

Murdock, G. P. (1965). Tenino Shamanism. Ethnology, 4, 165-171.

Murphy, J. M. (1964). Psychotherapeutic Aspects of Shamanism on St. Lawrence Island, Alaska. In Kiev, A. (Ed.), Magic, Faith and Healing Studies in Primitive Psychiatry Today (pp. 53-83). New York, NY: Free Press.

Musser Lopez, R. A. (1983). Yaa?vya's Poro: The Singular Power Object of a Chemehuevi Shaman. Journal of California and Great Basin Anthropology, 5, 260-264.

Myerhoff, B. G. (1966). The Doctor as Culture Hero: The Shaman of Rincon. Anthropological Quarterly, 39, 60-72.

Nadel, S. F. (1946). A Study of Shamanism in the Nuba Mountains. Journal of the Royal Anthropological Institute of Great Britain and Ireland, 76(1), 25-37.

Nappaaluk, S. M. (1997). Un temoignage inedit de Mitiarjuk sur les 'mumitsimajit' de Baie aux Feuilles, et sur les 'uirsaliit' et 'nuliarsaliit' du Nunavik. Etudes/Inuit/Studies, 21(1-2), 249-254.

Narby, J., & Huxley, F. (Eds.). (2001). Shamans Through Time: 500 Years on the Path to Knowledge. New York, NY: Jeremy P. Thacher/Putnam.

Nawa, K. (2004). Shaman Declaration. Jåohåo kanri = Journal of Information Processing and Management, 47(Part 1), 42-44.

Neimark, J. (1993). Shaman in Chicago. Psychology Today, 26(5), 46.

Neimark, J. (1996). Calling All Plant Spirits. New Age Journal, 13(4), 82.

Nencini, P. (2002). The Shaman and the Rave Party: Social Pharmacology of Ecstasy. Substance Use Misuse, 37(8-10), 923.

Neu, J. (1975). Levi-Strauss on Shamanism. Man, 10(2), 285-292.

Newcomb, F. J. (1964). Hosteen Klah, Navaho Medicine Man and Sand Painter. Norman, OK: University of Oklahoma Press.

Niquette, C. M., Konigsburg, L. W., & Hand, R. B. (1995). The Lead Branch Crematory (15PE126), Perry County, Kentucky. Midcontinental Journal of Archaeology, 20(2), 141-166.

Nishimura, K. (1987). Shamanism and Medical Cures. Current Anthropology, 28(4), S59-S64.

Noll, R. (1983). Shamanism and Schizophrenia: A State Specific Approach to the "Schizophrenic Metaphor" of Shamanistic States. American Ethnologist, 10, 443-459.

Noll, R. (1985). Mental Imagery Cultivation as a Cultural Phenomenon: The Role of Visions in Shamanism. Current Anthropology, 26(4), 443-461.

Noll, R. (1989). What Has Really Been Learned About Shamanism? Journal of Psychoactive Drugs, 21(1), 47-50.

Noll, R. (1990). Comment on 'Individuation and Shamanism'. Journal of Analytical Psychology, 35(2), 213.

Nomland, G. A. (1931). A Bear River Shaman's Curative Dance. American Anthropologist, 33, 38-41.

Oakes, M. (1978). The Blessing Way Ceremony. In Hall, G. (Ed.), The Shaman from Elko: Papers in Honor of Joseph L. Henderson on his Seventy-Fifth Birthday (pp. 59-63). San Francisco, CA: Jung Institute.

Ohlmarks, A. (1939). Studien zum Problem des Shamanismus. Lund: [s.n.].

Oleksa, M. J. (1977). Orthodoxy in Alaska: The Spiritual History of the Kodiak Aleut People. St. Vladimir's Theological Quarterly, 25(1), 3-19.

Olofson, H. (1979). Northern Paiute Shamanism Revisited. Anthropos, 74, 11-24.

Olson, R. L. (1961). Tlingit Shamanism and Sorcery. [s.n.]: Kroeber Anthropological Society.

O'Malley, L. D. (1997). The Monarch and the Mystic: Catherine the Great's Strategy of Audience Enlightenment in The Siberian Shaman. Slavic and East European Journal, 41(2), 224.

O'Neil, D. H. (1983). A Shaman's "Sucking Tube" from San Diego County, California. Journal of California and Great Basin Anthropology, 5, 245-247.

O'Neil, J. D. (1979). Illness in Inuit society: Traditional Context and Acculturative Influences. Na'Pao, 1(1-2), 40-50.

Oosten, J. G. (1981). The Structure of the Shamanistic Complex Among the Netsilik and Iglulik. Etudes/Inuit/Studies, 5(1), 83-98.

Oosten, J. G. (1986). Male and Female in Inuit Shamanism. Etudes/Inuit/Studies, 10(1-2), 115-131.

Oosten, J. G., & Remie, C. H. W. (1997). 'Angakkut' and Reproduction: Social and Symbolic Aspects of Netsilik Shamanism. Etudes/Inuit/Studies, 21(1-2), 75-100.

Opler, M. E. (1936). Some Points of Comparison and Contrast Between the Treatment of Functional Disorders by Apache Shamans and Modern Psychiatric Practice. American Journal of Psychiatry, 92, 1371-1387.

Opler, M. E. (1943). Navaho Shamanistic Practice Among the Jicarilla Apache. New Mexico Anthropologist, 6/7, 13-18.

Opler, M. E. (1946). The Creative Role of Shamanism in Mescalero Apache Mythology. Journal of American Folklore, 59, 268-281.

Opler, M.E. (1947). Notes on Chiricahua Apache Culture 1: Supernatural Power and Shaman. Primitive Man, 20(1-2), 1-14.

Opler, M. E. (1969). Apache Odyssey: A Journey Between Two Worlds. New York, NY: Holt, Rinehart and Winston.

Opler, M. E. (1985). The Creative Role of Shamanism in Mescalero Apache Mythology. In Ford, R. I. (Ed.), The Ethnographic American Southwest (pp. 268-281). New York, NY: Garland Publishing.

Oppitz, M. (1993). Who Heals the Healer? Shaman Practice in the Himalayas. Psychotherapy and Psychosomatic Medical Psychology, 43(11), 387-395.

Orozco Nunez, E. (1999). Comments on the Article "The Shamanic Cure: A Psychosocial Interpretation". Salud Publica Mexico, 41(4), 260.

Ott, J. (2001). Pharmanopo-Psychonautics: Human Intranasal, Sublingual, Intrarectal, Pulmonary and Oral Pharmacology of Bufotenine. Journal of Psychoactive Drugs, 33(3), 273-281.

Oubre, A. (1995). Social Context of Complementary Medicine in Western Society, Part I. Journal of Alternative and Complementary Medicine, 1(1), 41-56.

Oubre, A. Y., Carlson, T. J., King, S. R., & Reaven, G. M. (1997). From Plant to Patient: An Ethnomedical Approach to the Identification of New Drugs for the Treatment of NIDDM. Diabetologia, 40(5), 614-617.

Owens, R., & Sabina, M. (2004). Book Reviews - American Shaman - Selections. The American Book Review, 25(5), 1.

Oyuela-Caycedo, A. (2001). The Rise of Religious Routinization The Study of Changes from Shaman to Priestly Elite. BAR International Series (supplementary), 982, 5-18.

Paisano Suazo, A. (1979). Suggested Perspectives in Counseling the American Indian Client. [s.n.], NM: [s.n.].

Palmer, L. (1996). Walk on the Wild Side: Go to the Mountains with a Man Studying to Become a Shaman, a Messenger Between the Spirit World and the Ordinary World. Natural Health, 26(5), 86.

Park, S. (1986). Samson Grant: Atsuge Shaman. Redding, CA: Redding Museum and Art Center.

Park, W. Z. (1934). Paviotso Shamanism. American Anthropologist, 36, 98-113.

Park, W. Z. (1938). Shamanism in Western North America: A Study in Cultural Relationships. Evanston, IL: Northwestern University Press.

Park, W. Z. (1975). Shamanism in Western North America: A Study in Cultural Relationships: Reprint of 1938. New York, NY: Cooper Square Publishers.

Parker, A. C. (1928). Indian Medicine and Medicine Men. Toronto, Ontario.

Parker, A. C. (1974). Neh ho-noh-tci-noh-gah, the Guardians of the Little Waters, a Seneca Medicine Society. In Converse, H. M. (Ed.), Myths and Legends of the New York State Iroquois (pp. 149-183). Albany, NY: New York State Museum and Science Service.

Parker, A. C. (1985). Secret Medicine Societies of the Seneca. In Tooker, E. (Ed.), An Iroquois Source Book: Medicine Society Rituals (Vol. 3, pp. 161-185). New York, NY: Garland Publishing.

Parsons, E. C. (1985). Ceremonial Organization. In Ford, R. I. (Ed.), The Ethnographic American Southwest (pp. 108-167). New York, NY: Garland Publishing.

Pasztory, E. (1982). Shamanism and North American Indian Art. In Mathews Pearlstone, Z. & Jonaitis, A. (Eds.), Native North American Art History: Selected Readings (pp. 7-30). Palo Alto, CA: Peek Publications.

Paul, L. (1975). Recruitment to a Ritual Role. Ethos, 3(5), 449-467.

Paul, R. A. (1998). Shamans in History (Thomas and Humphrey's Shamanism, History, and the State, Humphrey and Onon's Shamans and Elders). Current Anthropology, 39(2), 2.

Pelcastre-Villafuerte, B. (1999). The Shamanic Cure: A Psychosocial Interpretation. Salud Publica Mexico, 41(3), 221-229.

Peters, Larry G., & Douglas Price-Williams. (1980). Towards an Experiential Analysis of Shamanism. American Ethnologist, 7(3), 397-413.

Peters, Larry G. (1982). Trance, Initiation, and Psychotherapy in Tamang Shamanism. American Ethnologist, 9(1), 21-46.

Phalon, R. (1994). Shaman Pharmaceuticals. Forbes, 153(8), 78.

Plotkin, M. (1996). Mark Plotkin, PhD: In Search of Plants that Heal. Interview by Bonnie Horrigan. Alternative Therapies in Health Medicine, 2(2), 66-75.

Plotnikoff, G. A., Numrich, C., Wu, C., Yang, D., & Xiong, P. (2002). Hmong Shamanism: Animist Spiritual Healing in Minnesota. Minnesota Medicine, 85(6), 7.

Plotnikoff, G. A., Numrich, C., Yang, D., Wu, C. Y., & Xiong, P. (2002). Shamans and Conventional Care: Are We Prepared? HEC Forum, 14(3), 271-278.

Polimeni, J., & Reiss, J. P. (2002). How Shamanism and Group Selection May Reveal the Origins of Schizophrenia. Medical Hypotheses, 58(3), 5.

Pond, G. H. (1854). Power and Influence of Dakota Medicine-Men. In Schoolcraft, H. R. (Ed.), Information Respecting the History, Condition, and Prospects of the Indian Tribes of the United States (Vol. 4, pp. 641-651). Philadelphia, PA: [s.n.].

Porterfield, A. (1984). Native American Shamanism and the American Mind-Cure Movement: A Comparative Study of Religious Healing. Horizons, 11(2), 276-289.

Porterfield, A. (1985). Algonquian Shamans and Puritan Saints. Horizons, 12(2), 303-310.

Porterfield, A. (1987). Shamanism: A Psychosocial Definition. Journal of the American Academy of Religion, 55(4), 721-739.

Porterfield, A. (1992). Witchcraft and the Colonization of Algonquian and Iroquois Cultures. Religion and American Culture, 2(1), 103-124.

Posinsky, S. H. (1965). Yurok Shamanism. Psychiatric Quarterly, 39, 227-243.

Powell, J. W. (1899). Eighteenth Annual Report of the Bureau of American Ethnology to the Secretary of the Smithsonian Institution, 1896-97, Part 1. Washington, DC: Government Printing Office.

Powers, W. K. (1986). Sacred Language: The Nature of Supernatural Discourse in Lakota. Norman, OK: University of Oklahoma Press.

Priestley, T. (1887). Report of Yakama Agent. In Annual Report of the Commissioner of Indian Affairs to the Secretary of the Interior for the Year 1887 (pp. 220-225). Washington, DC: Government Printing Office.

Prince, Raymond. (1982). Introduction to Shamans and Endorphins. Ethos, 10(4), 299-302.

Prince, Raymond. (1982). Shamans and Endorphins: Hypothesis for a Synthesis. Ethos, 10(4), 409-423.

Prucha, Z. S. (1979). The Virus and the Shaman: Must a GP Choose Between Truth and Consequences? Canadian Medical Assocication Journal, 121(6), 791.

Quinlan, A. R. (2000). The Ventriloquist's Dummy: A Critical Review of Shamanism and Rock Art in Far Western North America. Journal of California and Great Basin Anthropology, 22(1), 92-108.

Quinlan, A. R. (2000). Reply to Whitley. Journal of California and Great Basin Anthropology, 22(1), 129-132.

Rabinowitz, E. (2003). Care that Bridges Worlds. Healthplan, 44(4), 14-8.

Radin, P. (1911). The Ritual and Significance of the Winnebago Medicine Dance. The Journal of American Folklore, 24(92), 149-209.

Radin, P. (1914). Religion of the North American Indians. The Journal of American Folklore, 27(106), 335-373.

Radin, P. (1925). Thunder-Cloud, a Winnebago Shaman, Relates and Prays. In Parsons, E. C. (Ed.), American Indian Life (pp. 75-80). New York, NY: [s.n.].

Rain, M. S. (1985). Spirit Song: The Visionary Wisdom of No-eyes. Norfolk, VA: Donning Company.

Rain, M. S. (1987). Phoenix Rising: No-Eyes' Vision of the Changes to Come. Norfolk, VA: Donning Company.

Ramsey, J. (1989). The Poetry and Drama of Healing: The Iroquoian Condolence Ritual and the Navajo Night Chant. Literature and Medicine, 8, 78-99.

Rasmussen, K. (1975). A Shaman's Journey to the Sea Spirit Takanakapsaluk. In Tedlock, D. & Tedlock, B. (Eds.), Teachings from the American Earth (pp. 13-19). New York, NY: Liveright.

Ratner Shternberg, S. A. (1927). Muzeinye Materialy po Tlingitam. Paper presented at the Akademiia Nauk SSSR, Muzei Antropologii i Etnografii, Moscow, USSR.

Ray, V. F. (1936). The Kolaskin Cult: A Prophet Movement of 1870 in Northeastern Washington. American Anthropologist, 38(1), 67-75.

Razali, S. M. (1999). Conversion Disorder: A Case Report of Treatment with the Main Puteri, a Malay Shamanistic Healing Ceremony. European Psychiatry, 14, 470-472.

Reichard, G. A. (1934). Spider Woman: A Story of Navajo Weavers and Chanters. New York, NY: Macmillan.

Reichard, G. A. (1939). Navajo Medicine Man: Sandpaintings and Legends of Miguelito. New York, NY: J.J. Augustin.

Reichard, G. A. (1944). The Story of the Navajo Hail Chant. New York, NY: The Author.

Reichard, G. A. (1968). Spider Woman: A Story of Navajo Weavers and Chanters. Glorieta, NM: Rio Grande Press.

Riches, D. (1994). Shamanism: The Key to Religion. Man, 29(2), 381.

Riddell, F. A. (1955). Notes on Yokuts Weather Shamanism and the Rattlesnake Ceremony. Masterkey, 29, 94-98.

Ridington, R. (1987). From Hunt Chief to Prophet: Beaver Indian Dreamers and Christianity. Arctic Anthropology, 24(1), 8-18.

Ridington, R., & Ridington, T. (1970). Inner Eye of Shamanism and Totemism. History of Religions, 10, 49-61.

Ridington, R., & Ridington, T. (1975). The Inner Eye of Shamanism and Totemism. In Tedlock, D. & Tedlock, B. (Eds.), Teachings from the American Earth (pp. 190-204). New York, NY: Liveright.

Ripinsky-Naxon, M. (1993). The Nature of Shamanism. New York, NY: State University of New York Press.

Risse, G. B. (1972). Shamanism: The Dawn of a Healing Profession. Wisconsin Medical Journal, 71(12), 18-23.

Ritzenthaler, R. E., & Ritzenthaler, P. (1983). The Woodland Indians of the Western Great Lakes. Milwaukee, WI: Milwaukee Public Museum.

Robert Lamblin, J. (1997). Les Chamanes du Groenland Oriental: Elements Biographiques et Genealogiques. Etudes/Inuit/Studies, 21(1-2), 269-292.

Robertson, I. (1995). No Mana for the Shaman. BMJ: British Medical Journal, 311(7019), 1577.

Robinson, S. A. (1970). Comment on "Shamanism: The Beginnings of Art" by Andreas Lommel. Current Anthropology, 11, 45.

Rollmann, H. (1985). Inuit Shamanism and the Moravian Missionaries of Labrador: A Textual Agenda for the Study of Native Inuit Religion. Etudes/Inuit/Studies, 8(2), 131-138.

Romanucci-Ross, L. (1989). The Impassioned Cogito: Shaman and Anthropologist. In Hoppal, M. & Von Sadovszky, O. J. (Eds.), Shamanism: Past and Present (pp. 35-42). Budapest, Hungary: International Society for Trans-Oceanic Research.

Rose, W. (1984). Just What's All This Fuss About White Shamanism Anyway? In Scholer, B. (Ed.), Coyote Was Here (pp. 13-24). Aarhus, Denmark: Seklos, Department of English, University of Aarhus.

Russell, L. W. (1995). A Description of Some Interesting Artifacts. Bulletin of the Archaeological Society of Connecticut, 58, 3-12.

Saladin d'Anglure, B. (1983). Ijiqqat: Voyage au pays de l'invisible Inuit. Etudes/Inuit/Studies, 7(1), 67-83.

Saladin d'Anglure, B. (1986). Du foetus au Chamane: La Construction d'un "Troisieme Sexe" Inuit. Etudes/Inuit/Studies, 10(1-2), 25-113.

Saladin d'Anglure, B. (1988). Penser le Feminin Chamanique ou le Tiers-sex des Chamanes Inuit. Recherches Amerindiennes au Quebec, 18, 19-50.

Saladin d'Anglure, B. (1988). Kunut et les Angakkut Iglulik: Des Chamanes, Des Mythes et Des Tabous ou les Premiers Defis de Rasmussen en Terre Inuit Canadienne. Etudes/Inuit/Studies, 12(1-2), 57-80.

Saladin d'Anglure, B. (1990). Frerer-lune (Taqqiq), soeur-soleil (Siqiniq) et l'intelligence du Monde (Sila): Cosmologie Inuit, Cosmographie Arctique et Espace-temps Chamanique. Etudes/Inuit/Studies, 14(1-2), 75-139.

Saladin d'Anglure, B. (1994). From Foetus to Shaman: The Construction of an Inuit Sex. In Mills, N. & Slobodin, R. (Eds.), Amerindian Rebirth: Reincarnation Belief Among North American Indians and Inuit (pp. 82-106). Toronto, Canada: University of Toronto Press.

Saladin d'Anglure, B. (1997). Pour un Nouveau Regard Ethnographique le Chamanisme, La Possession et La Christianisation. Etudes/Inuit/Studies, 21(1-2), 5-20.

Saladin d'Anglure, B. (1997). A New Look on Shamanism, Possession, and Christianization. Etudes/Inuit/Studies, 21(1-2), 21-36.

Saladin d'Anglure, B. (2001). La Construction de L'identite Chamanique chez les Inuit du Nunavut et du Nunavik. Etudes/Inuit/Studies, 25(1-2), 191-215.

Saladin d'Anglure, B., & Hansen, K. G. (1997). Svend Frederiksen et le Chamanisme Inuit ou la Circulation des Noms (atiit), des Ames (tarniit), des Dons (tunijjutit) et des Espirits (tuurngait). Etudes/Inuit/Studies, 21(1-2), 37-73.

Saladin d'Anglure, B., & Philbert, J. (1993). The Shaman's Share, or Inuit Sexual Communism in the Canadian Central Arctic. Anthropologica, 35(1), 59-103.

Salzer, R. J. (1972). Bear-Walking: A Shamanistic Phenomenon Among the Potawatomi Indians in Wisconsin. Wisconsin Archeologist, 53, 110-146.

Sampath, H. M. (1988). Missionaries, Medicine and Shamanism in the Canadian Eastern Arctic. Arctic Medical Research, 47(1), 303-307.

Samuel, G. (1996). Nature Religion Today: Western Paganism, Shamanism and Esotericism in the 1990s Conference at the Lake District Campus of Lancaster University, 9th to 13th April 1996. Religion, 26(4), 4.

Sandner, D. F. (1978). The Navaho Prayer of Blessing. In Hall, G. (Ed.), The Shaman from Elko: Papers in Honor of Joseph L. Henderson on his Seventy-Fifth Birthday (pp. 35-38). San Francisco, CA: Jung Institute.

Sandner, D. F. (1979). Navajo Symbols of Healing. New York, NY: Harcourt Brace Jovanovich.

Sandner, D. F. (1979). Navaho Indian Medicine and Medicine Men. In Sobel, D. S. (Ed.), Ways of Health: Holistic Approaches to Ancient and Contemporary Medicine (pp. 117-146). New York, NY: Harcourt.

Sarris, G. (1992). Telling Dreams and Keeping Secrets: The Bole Maru as American Indian Religious Resistance. American Indian Culture and Research Journal, 16(1), 71-85.

Saunders, N. J. (1994). Predators of Culture: Jaguar Symbolism and Mesoamerican Elites. World Archaeology, 26(1), 104-117.

Scharfetter, C. (1985). The Shaman: Witness of an Old Culture—Is It Revivable? Schweizer Archiv fur Neurologie und Psychiatrie, 136(3), 81-95.

Schlesier, K. H. (1987). The Wolves of Heaven: Cheyenne Shamanism, Ceremonies, and Prehistoric Origins. Norman, OK: University of Oklahoma Press.

Schneider, G. W., & DeHaven, M. J. (2003). Revisiting the Navajo Way: Lessons for Contemporary Healing. Perspectives in Biological Medicine, 46(3), 413-427.

Seguin, M. (1984). The Tsimshian Images of the Past: Wiews for the Present. Vancouver, BC: University of British Columbia Press.

Seguin, M. (1984). Introduction. In Seguin, M. (Ed.), The Tsimshian (pp. ix-xx). Vancouver, BC: University of British Columbia Press.

Senn, C. F. (2002). Journeying as Religious Education: The Shaman, the Hero, the Pilgrim, and the Labyrinth Walker. Religious Education, 97(2), 17.

Senn, H. A. (1989). Jungian Shamanism. Journal of Psychoactive Drugs, 21(1), 113-121.

Shampo, M. A., & Kyle, R. A. (1982). The Shaman. Jama, 247(9), 1308.

Sheeley, W. F. (1962). A Brief History of Psychiatric Education for the Nonpsychiatrist. I. Shaman, Priest, Physician, Philosopher, Inquisitor. Psychosomatics, 3, 269-275.

Shimoji, A., Eguchi, S., Ishizuka, K., Cho, T., & Miyakawa, T. (1998). Mediation Between the Shamanistic Model and the Psychiatric Model in a Shamanistic Climate: A Viewpoint of Medical Anthropology. Psychiatry and Clinical Neuroscience, 52(6), 581-586.

Shimoji, A., & Miyakawa, T. (2000). Culture-bound Syndrome and a Culturally Sensitive Approach: From a Viewpoint of Medical Anthropology. Psychiatry and Clinical Neuroscience, 54(4), 461-466.

Shirokogoroff, S. (1935). Psychomental Complex of the Tungus. London, UK.

Silver, S. M., & Wilson, J. P. (1988). Native American Healing and Purification Rituals for War Stress. In Wilson, J. P., Harel, Z. & Kahana, B. (Eds.), Human Adaptation to Extreme Stress: From the Holocaust to Vietnam (pp. 337-355). New York, NY: Plenum Press.

Silverman, J. (1967). Shamans and Acute Schizophrenia. American Anthropologist, 69, 21-31.

Simms, S. C. (1906). The Metawin Society of the Bungees or Swampy Indians of Lake Winnipeg. Journal of American Folklore, 19, 330-333.

Sinnott, P. B. (1873). No. 73: Annual Report of Grand Ronde Agency. In Annual Report of the Commissioner of Indian Affairs to the Secretary of the Interior, for the Year 1873 (pp. 320-322). Washington, DC: Government Printing Office.

Siskin, E. E. (1983). Washo Shamans and Peyotists: Religious Conflict in an American Indian Tribe. Salt Lake City, UT: University of Utah Press.

Siskin, E. E. (1984). Rejoinder to Stewart. Reviews in Anthropology, 11, 341-344.

Skinner, A. B. (1985). Some Seneca Masks and Their Uses. In Tooker, E. (Ed.), An Iroquois Source Book: Medicine Society Rituals (pp. 191-207). New York, NY: Garland Publishing.

Smith, D. (1985). Witchcraft and Demonism of the Modern Iroquois. In Tooker, E. (Ed.), An Iroquois Sourcebook: Calendric Rituals (pp. 184-194). New York, NY: Garland Publishing.

Smith, D. (1985). Additional Notes on Onondaga Witchcraft and Hon-do'-i. In Tooker, E. (Ed.), An Iroquois Sourcebook: Calendric Rituals (pp. 277-281). New York, NY: Garland Publishing.

Smith, H. I. (1896). Certain Shamanistic Ceremonies among the Ojibwas. American Antiquarian and Oriental Journal, 18, 282-284.

Smith, M. W. (1954). Shamanism in the Shaker Religion of Northwest America. Man, 54, 119-122.

Smith, M. W. (1963). Strange Tales of Abenaki Shamanism. Lewiston, ME: Central Maine Press.

Smith, N. N. (1977). The Changing Role of the Wabanaki Chief and Shaman. In Cowan, W. (Ed.), Actes du Huitieme Congres des Algonquinistes (pp. 213-221). Ottawa, Canada: Carleton University.

Smith, N. N., & Walker, W. (1997). The Changing Role of Shamans and Their Magic in the Validation and Maintenance of Wabanaki Culture. In Pentland, D. (Ed.), Papers of the Twenty-Eighth Algonquian Conference (pp. 365-372). Winnipeg, Manitoba: Winnipeg University of Manitoba.

Smyers, K. A. (2001). Shaman/Scientist: Jungian Insights for the Anthropological Study of Religion. Ethos, 29(4), 16.

Soby, R. M. (1997). Naming and Christianity. Etudes/Inuit/Studies, 21(1-2), 293-301.

Sokolova, Z. P. (1989). A Survey of the Ob-Ugrian Shamanism. In Hoppal, M. & Von Sadovszky, O. J. (Eds.), Shamanism: Past and Present (pp. 155-164). Budapest, Hungary: International Society for Trans-Oceanic Research.

Speck, F. G. (1919). Penobscot Shamanism. Menasha, WI: American Anthropological Association.

Speck, F. G. (1985). How the Dew Eagle Society of the Allegany Seneca Cured Gahehdagowa (FGS). In Tooker, E. (Ed.), An Iroquois Source Book: Medicine Society Rituals (pp. 39-59). New York, NY: Garland Publishing.

Speck, F. G. (1987). The Creek Indians of Taskigi Town. In Sturtevant, W. C. (Ed.), A Creek Source Book (pp. 99-164). New York, NY: Garland Publishing.

Stead, R., & Oliver, K. (1987). Traditional Lakota Religion in Modern Life. In DeMallie, R. J. (Ed.), Sioux Indian Religion (pp. 211-216). Norman, OK: University of Oklahoma Press.

Steinbring, J. (1982). Shamanistic Manipulation and the Algonkian Idiom in the Archaeoethnology of Rock Art. American Indian Rock Art, 7-8, 212-220.

Stephen, M., & Suryani, L. K. (2000). Shamanism, Psychosis and Autonomous Imagination. Culture, Medicine and Psychiatry, 24(1), 34.

Stewart, K. M. (1970). Mojave Indian Shamanism. Masterkey, 44, 15-24.

Stewart, K. M. (1974). Mohave Shamanistic Specialists. Masterkey, 48, 4-13.

Stewart, O. C. (1956). Three Gods for Joe. Tomorrow, 4(3), 71-76.

Stewart, O. C. (1984). Siskin's Report did not Improve with Age. Reviews in Anthropology, 11, 144-157.

Stockel, H. H. (1993). Survival of the Spirit: Chiricahua Apaches in Captivity. Reno, NV: University of Nevada Press.

Stoney, S. A. (1994). Rock Rrt in the Great Basin: The Scratched Style Mystery Reexamined: Is it Illusion or Reality? Pacific Coast Archaeological Society Quarterly, 30(4), 33-54.

Storck, M., Csordas, T. J., & Strauss, M. (2000). Depressive Illness and Navajo Healing. Medical Anthropology Quarterly, 14(4), 571-597.

Storl, W. D. (1983). Die Indianer der Spiritisten. Wiener Ethnohistorische Blatter, 26, 3-18.

Stott, J. C. (1985). Spirits in the Snowhouse: The Inuit Angakok (Shaman) in Children's Literature. The Canadian Journal of Native Studies, 5(2), 193-200.

Straight, W. M. (1970). Josie Billie, Seminole Doctor, Medicine Man, and Baptist Preacher. Journal of the Florida Medical Association, 57, 33-40.

Stutley, M. (2003). Shamanism. TLS, The Times Literary Supplement, (5209), 30.

Suddick, R. P., & Harris, N. O. (1990). Historical Perspectives of Oral Biology: A Series. Critical Review of Oral Biological Medicine, 1(2), 135-151.

Sullivan, L. E. (1994). The Attributes and Power of the Shaman: A General Description of the Ecstatic Care of the Soul. In Seaman, G. & Day, J. S. (Eds.), Ancient Traditions: Shamanism in Central Asia and the Americas. Niwot, CO: University of Colorado Press.

Summer Rain, M. (1985). Spirit Song: The Visionary Wisdom of No-eyes. Norfolk, VA: Donning Company.

Sword, G. (1980). Treating the Sick [August 5, 1901]. In Walker, J. R. (Ed.), Lakota Belief and Ritual (pp. 91-93). Lincoln, NE: University of Nebraska Press.

Sword, G., Wound, B., Flesh, N., & Tyon, T. (1980). The Secret Knowledge of Shamans. In Walker, J. R. (Ed.), Lakota Belief and Ritual (pp. 93-96). Lincoln, NE: University of Nebraska Press.

Tacon, P. S. C. (1983). An Analysis of Dorset Art in Relation to Prehistoric Culture Stress. Etudes/Inuit/Studies, 7(1), 41-65.

Taegel, W. (1992). Psychotherapist and Shaman. Voices: The Art and Science of Psychotherapy, 28(4), 2.

Tanner, H. H. (1979). Coocoochee: Mohawk Medicine Women. American Indian Culture and Research Journal, 3(3), 23-41.

Tantaquidgeon, G. (1972). Folk Medicine of the Delaware and Related Algonkian Indians. Harrisburg, PA: Pennsylvania Historical and Museum Commission.

Tapatai, L., & Tapatai, L. (1972). Akulak, the Shaman. Inuttuttut; 1972. p. 7-11, 7-11.

Taylor, E. (1999). Shadow Culture: Psychology and Spirituality in America. Washington, D.C.: Counterpoint.

Taylor, E., & Pledilato, J. (2002). Shamanism and the American Psychotherapeutic Counter-Culture. Journal of Ritual Studies, 16(2), 129-140.

Taylor, J. G. (1997). Deconstructing Deities: 'Tuurngatsuak' and 'Tuurngaatsuk' in Labrador Inuit Religion. Etudes/Inuit/Studies, 21(1-2), 141-158.

Taylor, W. E., Jr. (1975). Speculations and Hypotheses on Shamanism in the Dorset Culture of Arctic Canada. In Centro, D. (Ed.), Valmonica Symposium 1972: Acted du Symposium International Sur les Religions de la Prehistoire (pp. 473-482). Capo Di Ponte: [s.n.].

Tedlock, D., & Tedlock, B. (Eds.). (1975). Teachings from the American Earth: Indian Religion and Philosophy. New York, NY: Liveright.

Tedlock, D., & Tedlock, B. (1975). Introduction. In Tedlock, D. & Tedlock, B. (Eds.), Teachings from the American Earth (pp. xi-xxiv). New York, NY: Liveright.

Tedlock, D., & Tedlock, B. (Eds.). (1978). Uber den Rand des Tiefan Canyon: Lehren indianischer Schamanen. Dusseldorf: Diedrichs.

Tens, I., & Barbeau, C. M. (1975). The Career of a Medicine-man. In Tedlock, D. & Tedlock, B. (Eds.), Teachings from the American Earth: Indian Religion and Philosophy (pp. 3-12). New York, NY: Liveright.

Thalbitzer, W. C. (1930). Les Magiciens Esquimaux, Leurs Conceptions du Monde, de L'ame et de la Vie. Societe des Americanistes, 22, 73-106.

Therrien, M. (1983). L'Inuk tel que le Revele le Langage Sacre: Quelques Observations. Inter-Nord, 17, 111-114.

Thomas, Humphrey, & Valeri. (1995). Shamanism, History, & the State. American Ethnologist, 22(3), 1.

Thomas, D. H. (1976). A Diegueno Shaman's Wand: An Object Lesson Illustrating the "Heirloom Hypothesis". Journal of California Anthropology, 3(1), 128-132.

Thomas, H., & Allan, S. (1997). Shamanism, History, and the State. The Journal of Asian studies, 56(3), 2.

Thompson, C. (1997). Structure, Metaphor, and Iconicity in Koyukon Shamanistic Stories. American Indian Quarterly, 21(2), 149-169.

Thomson, K., & Long, W. W. (1979). Curing Chants in the Notebooks of Two Cherokee Doctors. In Brotherston, G. (Ed.), Image of the New World: The American Continent Portrayed in Native Texts (pp. 253-254). London, UK: Thames and Hudson.

Thunderhorse, I. (1990). Return of the Thunderbeings: A New Paradign of Ancient Shamanism. Santa Fe, NM: Bear & Company.

Thurston, B. P. (1933). A Night in a Maidu Shaman's House. Masterkey, 7, 111-115.

Timothy, R. M. (1999). Confessions of a Doubtful Shaman. Campus-Wide Information Systems, 16(1), 39.

Titiev, M. (1956). Shamans, Witches and Chiefs Among the Hopi. Tomorrow, 4(3), 51-56.

Tompkins, P. (2001). What Kind of Errand?—Eccentricities of a Shaman. Parabola, 26(3), 5.

Tooker, E. (Ed.). (1979). Native North American Spirituality of the Eastern Woodlands: Sacred Myths, Dreams, Visions, Speeches, Healing Formulas, Rituals and Ceremonies. New York, NY: Paulist Press.

Tooker, E. J. (Ed.). (1985). An Iroquois Source Book. Vol. 3. Medicine Society Rituals. New York, NY: Garland Publishing.

Tooker, E. J. (1985). Introduction. In Tooker, E. (Ed.), An Iroquois Source Book: Medicine Society Rituals (pp. xv-xviii). New York, NY: Garland Publishing.

Topper, M. D. (1987). The Traditional Navajo Medicine Man Therapist, Counselor, and Community Leader. Journal of Psychoanalytic Anthropology, 10, 217-249.

Topper, M. D., & Schoepfle, G. M. (1990). Becoming a Medicine Man: A Means to Successful Midlife Transition Among Traditional Navajo Men. In Nemiroff, R. A. & Calarusso, C. A. (Eds.), New Dimensions in Adult Development (pp. 443-466). New York, NY: Basic Books.

Torrey, E. Fuller. (1974). Spiritualists and Shamans as Psychtherapists: An Account of Original Anthropological Sin. In I.I. Zaretsky, and M.P. Leone, eds., Religious Movements in Contemporary America (pp. 330-337). Princeton, N.J.: Princeton University Press.

Trigger, B. G. (1981). Ontario Native People and the Epidemics of 1634-1640. In Krech, S. I. (Ed.), Indians, Animals, and the Fur Trade: A Critique of Keepers of the Game (pp. 19-38). Athens, GA: University of Georgia Press.

Trott, C. G. (1997). The Rapture and the Rupture: Religious Change Amongst the Inuit of North Baffin Island. Etudes/Inuit/Studies, 21(1-2), 209-228.

True, D. L. (1986). To-vah: A Luiseno Power Cave. Journal of California and Great Basin Anthropology, 8, 269-273.

Tubby, E., & Tubby, V. (1976). Estelline Tubby Remembers. Nanih Waiya, 4(1), 15-18.

Tucker, M. (1998). Halloween Mask. In Andrews, S. B. & Creed, J. (Eds.), Authentic Alaska: Voices of its Native Writers (pp. 67-69). Lincoln, NE: University of Nebraska Press.

Tuohy, D. R., & Stein, M. C. (1969). A Late Lovelock Shaman and His Grave Goods.Unpublished manuscript, Carson City, NV.

Turgeon, L. (1997). The Tale of the Kettle: Odyssey of an Intercultural Object. Ethnohistory, 44(1), 1-29.

Turner, E. (1989). From Shamans to Healers: The Survival of an Inupiaq Eskimo Skill. Anthropologica, 31(1), 3-24.

Turner, E. (1996). The Hands Feel It: Healing and Spirit Presence Among a Northern Alaskan People. DeKalb, IL: Northern Illinois University Press.

Turner, E. (2004). Shamanism and Spirit. Expedition, 46(1), 4.

Turner, T., & Yanomami, D. K. (1991). "I Fight Because I Am Alive". Cultural Survival Quarterly, 15(3), 59.

Turpin, S. A. (1996). Painting on Bones and Other Unusual Media in the Lower and Trans-Pecos Region of Texas and Coahuila. Plains Anthropologist, 41(157), 261-272.

Tyon. (1995). The Number Four and the Circle. In Hirschfelder, A. (Ed.), Native Heritage: Personal Accounts by American Indians, 1790 to the Present (pp. 202). New York, NY: Macmillan.

Tyon, T., DeMallie, R. J., & Jahner, E. A. (1980). Bears are 'Wakan' ('Mato Wicayuwakanpi kin'). In Walker, J. R. (Ed.), Lakota Belief and Ritual (pp. 157-159). Lincoln, NE: University of Nebraska Press.

Tyon, T., DeMallie, R. J., & Jahner, E. A. (1980). Toads are 'Wakan' ('Witapir'a Yuwakanpi kin'). In Walker, J. R. (Ed.), Lakota Belief and Ritual (pp. 161). Lincoln, NE: University of Nebraska Press.

Underhill, R. M. (1946). Papago Indian Religion. New York, NY: Columbia University Press.

Vaella, Y. (2002). From Ischetka: The Little Shaman and Other Stories. The North Dakota Quarterly, 69(Part 3), 66.

Van Blerkom, L. M. (1995). Clown Doctors: Shaman Healers of Western Medicine. Medical Anthropology Quarterly, 9(4), 462-475.

Vastokas, J. M. (1973). The Shamanistic Tree of Life. Arts Canada, 30, 125-149.

Vaught, R., & Tippett, J. (2001). Control Performance Monitoring: Shaman or Saviour—A Candid Look at What Such a System Can Truly Do. Pulp & Paper Canada, 102(9), 4.

Vazeilles, D. (1996). Chamanes et Visionnaires Sioux. Monaco: Editions du Rocher/Le Mail.

Vitebsky, P. (1997). What Is a Shaman? Natural History, 106(2), 2.

Vogel, K. (2003). Female Shamanism, Goddess Cultures, and Psychedelics. Re-vision, 25(Part 3), 18-29.

Voght, M. (1991). Shamans and Padres: The Religion of the Southern California Mission Indians. In Castillo, E. D. (Ed.), Native American Perspectives on the Hispanic Colonization of Alta California. New York, NY: Garland Publishing.

von Keitz, E. (1999). Medicine Men. Journal of Christian Nursing, 16(4), 26-27.

von Stuckard, K. (2002). Reenchanting Nature: Modern Western Shamanism and Nineteenth-century Thought. Journal of the American Academy of Religion, 70(4), 771-800.

Voss, R. W., Victor Douville, A. L. S., & Gayla, T. (1999). Tribal and Shamanic-Based Social Work Practice: A Lakota Perspective. Social Work, 44(3), 228-241.

Waite, D. (1982). Kwakiutl Transformation Masks. In Mathews Pearlstone, Z. & Jonaitis, A. (Eds.), Native North American Art History: Selected Readings (pp. 137-155). Palo Alto, CA: Peek Publications.

Walker, D. E., Jr. (Ed.). (1989). Witchcraft and Sorcery of the American Native Peoples. Moscow, ID: University of Idaho Press.

Walker, J. R. (1980). Lakota Belief and Ritual. Lincoln, NE: University of Nebraska Press.

Walker, W. (1981). Cherokee Curing and Conjuring, Identity, and the Southeastern Co-tradition. In Pierre, G. & Kushner, G. (Eds.), Persistent Peoples: Cultural Enclaves in Perspective (pp. 86-105). Tucson, AZ: University of Arizona Press.

Wallis, R. (2000). Queer Shamans: Autoarchaeology and Neo-shamanism. World Archaeology, 32(2), 252-262.

Walraven, B., & Howard, K. (1996). Songs of the Shaman. Bulletin of the School of Oriental and African Studies, University of London, 59(1), 1.

Walsh, R. (1994). The Making of a Shaman: Calling, Training, and Culmination. The Journal of Humanistic Psychology, 34(3), 7.

Wardwell, A. (1993). Some Discoveries in Northwest Coast Indian Art. American Indian Art Magazine, 18(2), 46-55.

Wardwell, A. (1996). Tangible Visions: Northwest Coast Indian Shamanism and its Art. New York, NY: Monacelli Press.

Warner, R. (1976). Deception in Shamanism and Psychiatry. Transnational Mental Health Research Newsletter, 18(3), 2.

Warner, R. (1980). Deception and Self-deception in Shamanism and Psychiatry. International Journal of Social Psychiatry, 26(1), 41-52.

Wax, M. L. (1995). Method as Madness Science, Hermeneutics, and Art in Psychoanalysis. Journal of the American Acadamy of Psychoanalysis, 23(4), 525-543.

Weathers, D. (1998). Culture: The Wisdom of the Somes. Essence, 29(8), 124.

Webber, A. P. (1983). Ceremonial Robes of the Montagnais-Naskapi. American Indian Art, 9(1), 60-69.

Wellmann, K. F. (1976). Schamanistische Bezuge in Nordamerikanischen Indianischen Felshildern. Ethnologia Americana, 15, 833-839.

Wellmann, K. F. (1981). Rock Art, Shamans, Phosphenes and Hallucinogens in North America. Bollettino del Centro Camuno di Studii Preistorici, 18, 89-103.

Wells, R., & Sheldon, M. (2000). Making Room for Alternatives. Hastings Center Rep, 30(3), 26-28.

Weslager, C. A. (1973). Magic Medicines of the Indians. Somerset, NJ: Middle Atlantic Press.

White, L. A. (1928). A Comparative Study of Keresan Medicine Societies. Paper presented at the International Congress of Americanists.

White, L. A. (1985). A Comparative Study of Keresan Medicine Societies. In Ford, R. I. (Ed.), The Ethnographic American Soutwest (pp. 604-619). New York, NY: Garland Publishing.

Whitehead, H. (2000). The Hunt for Quesalid: Tracking Levi-Strauss' Shaman. Anthropology & Medicine, 7(2), 149-168.

Whitley, D. S. (2000). Response to Quinlan. Journal of California and Great Basin Anthropology, 22(1), 108-129.

Wiek, S. (1977). Castaneda: Coming of Age in Sonora. American Anthropologist, 79, 84-91.

Wiercinski, A. (1989). On the Origin of Shamanism. In Hoppal, M. & Von Sadovszky, O. J. (Eds.), Shamanism: Past and Present (Vol. 1, pp. 19-25). Budapest, Hungary: International Society for Trans-Oceanic Research.

Wihr, W. S. (1995). "You Toad-sucking Fool": An Inquiry into the Possible Use of Bufotenine by Northern Northwest Coast Shamans. Northwest Anthropological Research Notes, 29(1), 51-59.

Wilk, S. (1978). On the Experiential Approach in Anthropology: A Reply to Maquet. American Anthropologist, 80, 363-364.

Williams, E. A. (1993). Gloria Flaherty Shamanism and the Eighteenth Century. The American historical review, 98(3), 859.

Willis, R. (1994). New Shamanism. Royal Anthropological Institute news : RAIN, 10(6), 16.

Wilson, B., & Hills, C. L. (1968). Ukiah Valley Pomo Religious Life, Supernatural Doctoring, and Beliefs: Observations of 1939-1941. Berkeley, CA: University of California Press.

Winkelman, M. (1984). A Cross-Cultural Study of Magico-Religious Practitioners. In Heinze, R.-I. (Ed.), Proceedings of the International Conference on Shamanism (Vol. 27-38): Berkeley: Independent Scholars of Asia.

Winkelman, M. (1989). A Cross-cultural Study of Shamanistic Healers. Journal of Psychoactive Drugs, 21(1), 17-24.

Winkelman, M. (1991). Physiological and Therapeutic Aspects of Shamanistic Healing. Subtle Energies, 1(2):1-18.

Winkelman, M. (1995). Psychointegrator Plants: Their Roles in Human Culture, Consciousness and Health. Yearbook of Cross-Cultural Medicine and Psychotherapy, 9-53.

Winkelman, M. (1996). Shamanism and Consciousness: Metaphorical, Political and Neurophenomenological Perspectives. Transcultural Psychiatric Research Review, 33(1), 69.

Winkelman, M. (2000). Shamanism: The Neural Ecology of Consciousness and Healing. Westport, CT: Bergin & Garvey.

Winkelman, M. (2001). Alternative and Traditional Medicine Approaches for Substance Abuse Programs: A Shamanic Perspective, International Journal of Drug Policy, 12, 337-351.

Winkelman, M. (2002). Shamanism and Cognitive Evolution (with comments). Cambridge Archaeological Journal, 12(1), 31.

Winkelman, M. (2002). Shamanism as Neurotheology and Evolutionary Psychology. American Behavioral Scientist, 45(12), 1875-1887.

Winkelman, M. (2003). Complementary Therapy for Addiction: "Drumming Out Drugs". American Journal of Public Health, 93(4), 647-51.

Winkelman, M. (2004). Shamanism as the Original Neurotheology. Zygon, 39(1), 193-217.

Winter, J. C. (2000). Traditional Uses of tobacco by Native Americans. In Winter, J. C. (Ed.), Tobacco Use by Native North Americans: Sacred Smoke and Silent Killer (pp. 9-58). Norman, OK: University of Oklahoma Press.

Winter, J. C. (2000). From Earth Mother to Snake Woman: The Role of Tobacco in the Evolution of Native American Religious Organization. In Winter, J. C. (Ed.), Tobacco Use by Native North Americans: Sacred Smoke and Silent Killer (pp. 265-304). Norman, OK: University of Oklahoma Press.

Wissler, C. (1912). Societies of the Plains Indians. New York, NY: The Trustees.

Wissler, C. (1916). General Discussion of Shamanistic and Dancing Societies. New York, NY: American Museum of Natural History.

Wissler, C. (1925). Smoking Star, a Blackfoot Shaman. In Parsons, E. C. (Ed.), American Indian Life (pp. 45-62). New York, NY: [s.n.].

Witthoft, J. (1983). Cherokee Beliefs Concerning Death. Journal of Cherokee Studies, 8, 68-72.

Wittmer, M. K. (1981). Art and Shamanism. Native Arts/West, 1(9), 6-11.

Wolf. (1990). The Woman Who Didn't Become a Shaman. American Ethnologist, 17(3), 419-430.

Wolf, V. (1994). The Eagle Site. Pacific Coast Archaeological Society Quarterly, 30(4), 55-71.

Wooden, A. C. (1958). Umigiukor; the Shaman. Medical Times, 86(11), 1407-18.

Woon, T. H., & Teoh, C. L. (1976). Psychotherapeutic Management of a Potential Spirit Medium. Australian and New Zeland Journal of Psychiatry, 10(1A), 125-8.

Wootton, D., & Behringer, W. (1999). Shaman of Oberstdorf. The London Review of Books, 21(22), 34.

Wound, L., Horse, A., Star, L., & Herman, A. (1980). Instructing Walker as a Medicine Man. By Little Wound, American Horse, and Lone Star, Interpreted by Antoine Herman [September 12, 1896]. In Walker, J. R. (Ed.), Lakota Belief and Ritual (pp. 68). Lincoln, NE: University of Nebraska Press.

Wright, A. R. (1977). First Medicine Man: The Tale of Yobaghu-Talyonunh. Anchorage, AK: O.W. Frost.

Wright, P. A. (1995). The Interconnectivity of Mind, Brain, and Behavior in Altered States of Consciousness: Focus on Shamanism. Alternative Therapies in Health and Medicine, 1(3), 50-56.

Wyman, L. C. (1983). Southwest Indian Drypainting. Santa Fe, NM: School of American Research.

Yamada, T. (1996). Through Dialogue with Contemporary Yakut Shamans: How They Revive Their Worldview. Anthropology of Consciousness, 7(3), 1-14.

Yellowtail, T., & Fitzgerald, M. O. (1991). Yellowtail, Crow Medicine Man and Sun Dance Chief: An Autobiography. Norman, OK: University of Oklahoma Press.

York, M. (2001). New Age Commodification and Appropriation of Spirituality. Journal of Contemporary Religion, 16(3), 361-372.

Young, D., Ingram, G., & Swartz, L. (1989). Cry of the Eagle: Encounters with a Cree Healer. Toronto, Ontario: University of Toronto Press.

Zatzick, D. F., & Johnson, F. A. (1997). Alternative Psychotherapeutic Practice Among Middle Class Americans: II: Some Conceptual and Practical Comparisons. Culture, Medicine, and Psychiatry, 21(2), 213-246.

Zelizer, B. (1992). On Communicative Practice: The "Other Worlds" of Journalism and Shamanism. Southern Folklore, 49(1), 19-36.

Zigmond, M. L. (1977). The Supernatural World of the Kawaiisu. In Blackburn, T. C. (Ed.), Flowers of the Wind (pp. 59-95). Socorro, NM: Ballena Press.

Zolla, E. (1973). The Writer and the Shaman: A Morphology of the American Indian. New York, NY: Harcourt Brace Jovanovich.

www.ingramcontent.com/pod-product-compliance
Lightning Source LLC
Chambersburg PA
CBHW051734090426
42738CB00010B/2248